MONSTER MOVES

MONSTER MOVES

ADVENTURES MOVING THE WORLD'S BIGGEST STRUCTURES

CARLO MASSARELLA

FIREFLY BOOKS

Contents

Introduction

This book is based on the adventures we have experienced and untold stories we have encountered in making the television series *Monster Moves*. It charts the daring exploits of transport specialists as they attempt to move immense buildings, machines and vehicles across the globe, by land, sea and air.

An eye-catching photograph that landed in my inbox one morning inspired the idea for *Monster Moves*. The picture depicted a massive house, balanced boldly on trucks and wheels, crossing over a highway bridge. The quest to verify whether the image was real or not opened my eyes to a universe of astonishing stories and compelling characters. Generations of "house-moving" families had, it appeared, mastered the seemingly impossible craft of relocating structures that were not built to leave their foundations. No edifice seemed too large for them to transport and the motivation for each move was always intriguing. Structural movers fearlessly rolled lighthouses off the edges of crumbling cliffs, motored mansions out of flood plains, and hauled entire towns to prevent them falling into chasms.

Fellow producer Jamie Lochhead and myself soon found ourselves in hot pursuit of our first building. House mover Jimmy Matyiko invited us to Virginia Beach to film him haul a beach house up the sands to a new site inland. Things didn't go quite as we had expected. With the house hooked to a truck, Jimmy drove it up the beach, playfully dodging the incoming waves and sunbathers in his path. Then, over the course of several hours, he skillfully steered it up the coast, onto the main road and under rows of low-slung power lines. Jamie and I stumbled behind, dazed by this improbable sight and exhausted by the heat. We had to abandon our car, stuck firm in the sand. We lost our shoes, tripped over dunes and smashed our camera equipment while sprinting to keep up. As we sat collapsed on the roadside in a puddle of sweat, we watched as the house disappeared over the horizon. The building made it to its new home unscathed, to the delight of its owners. We, however, were left battered and bone-tired. The move had been a spectacle to behold and we were determined to find a way to convey the thrill of this unique chase to a wider audience.

This book recounts some of the most unusual and memorable moves we have followed and goes behind the scenes to meet the brains and brawn behind each project. The movers themselves are a fearless band of hard-working, ingenious adventurers. These are the people I've had the great pleasure of spending much of the past decade with and I am truly in awe of their skills and talent. We also meet the individuals, families and communities who own the buildings and machines on the move. Together, their testimonies tell us something profound about our relationship with the structures we build, inhabit and maintain.

The book also explores histories that we have unearthed but not been able to feature in the TV series. Many of these accounts have proved hugely inspirational for us. They include the tale of the Hodge family from Suffolk, England, who took the courageous decision to relocate their Elizabethan mansion across the countryside to retain its stunning views, the remarkable case of Electra Webb, who masterminded the "sailing" of a colossal steamship across land, and the incredible feats of John Eichleay Jr., whose ambitious schemes to raise mansions up cliffs and leapfrog houses over trees were far ahead of their time.

A loyal group of dedicated people has worked with me over the years to chronicle the journeys of these supersize loads. Their research and dedication in the field have enabled us to draw together this rich tapestry of tales. I'm particularly indebted to the following producers who have made significant contributions to this book, relating their own experiences on location to bring the stories vividly to life: Leesa Rumley (Massive Monuments), Jamie Lochhead (The Floating Town), William Lorimer (Saving Sankaty Head Lighthouse), John Driftmier (The *Onondaga Saga*: Moving Canada's Submarine), Lee Reading (Spectacular Spitfire: The Flying Jigsaw) and Cherry Dorrett (The Man Who Collected a Town).

We hope you enjoy the journey as much as we have.

Carlo Massarella
Creator and Producer, *Monster Moves*

Mammoth Mansions

MOVING
THE FAMILY ESTATE

Packing up and moving house is a logistics nightmare. But imagine the daunting challenge of uprooting and moving your *entire* house intact – complete with walls, roof, chimneys and furnishings – to a new setting. This was the only way that Rose Heet could save her historic farm estate from the wrecking ball.

MOVE STATISTICS
Rose Heet's Farmhouse

Built	1890
Material	Brick
Weight	428 tons
Height	30 ft. (9 m)
Length	61 ft. (18.5 m)
Width	41 ft. (12.5 m)
Distance moved	1,600 ft. (500 m)
Number of wheels	112

Rose Heet's family had occupied their farm estate in Belleville on the outskirts of St. Louis, Missouri, for over 100 years. Their white, two-story, brick farmhouse was completed in 1890 by Rose's great-grandfather, Peter Voellinger. His initials together with the year the house was built, "PV 1890," were spelled out in a red-and-black patchwork of tiles on its rooftop. Several outbuildings, including a machine shop, a barn, a milk house and an outdoor toilet, were erected on the estate shortly after the main house was built.

These period buildings were more than just bricks and mortar to the Heet family. "It was passed down through the generations, from my great-great-grandmother, to my great-grandparents, to my grandparents, to my mother and my uncle and finally to me. It's very much more than just a piece of property, it's family," explains Rose.

Fifty years ago, a highway was built across the family farm that turned their land into prime real estate. For three decades, Rose had been fighting off developers who wanted to build a shopping center on her family's home. In 2004, in a final attempt to persuade her, developers offered to buy a lucrative piece of the farm standing on the highway, but to also cover the cost of moving Rose's historic farmstead to a more secluded section of her land 1,600 feet (500 m) to the east.

The family faced an agonizing decision. Stay put, but continue to battle developers, or uproot the entire farmstead, taking it off the much-prized land but preserving it for future generations of the family?

"I guess the debate was if the opportunity arose that we could move the homestead and preserve it then that was the decision that needed to be made," recalls Rose.

Top The Voellinger family in front of the farmhouse in 1908

Above The farmhouse before relocation in 2006

HISTORIC MOVES

The idea of moving an entire farmstead of structures from one location to another may seem outlandish, but Americans have been relocating buildings successfully since the early 18th century. There are many reasons to move a building. Sometimes a historically important building, such as Rose's farmstead, is threatened with demolition so that the land it is sitting on can be redeveloped. Moving the building onto a less lucrative plot offers a lifeline to its owners.

Sometimes a building sitting on unstable ground is in danger of sinking or collapse. Lighthouses perched on the edges of crumbling cliff tops often need to be hauled inland to prevent them from falling into the sea. Often homeowners living on constricted plots move their houses *vertically* to squeeze in another floor underneath. Occasionally, entire towns built on flood plains are moved to higher ground to protect them from deluges.

Moving any structure is a precarious and highly skilled operation. In North America, there are over 200 professional "structural movers," or "house movers," who advertise their services in the yellow pages, alongside regular plumbers, decorators and electricians. Many are family-run businesses that have been operating for generations.

Rose Heet called upon the services of John Matyiko from Expert House Movers, Missouri, to move her farmstead. The Matyiko family has forged an unrivaled reputation in America for moving immense structures that were never designed to leave their foundations. With offices in Maryland, Missouri and Virginia, the family played a key role in moving Sankaty Lighthouse on Nantucket and Cape Hatteras Lighthouse in North Carolina.

Brick buildings are by far the most challenging structures to move. Extremely heavy and fragile, they present the kind of task that would keep most of us awake at night. But John's easygoing sense of humor and confident manner quickly reassured Rose that her farmstead would be in safe hands, even though it would be a formidable challenge for John's seasoned team.

"It's a masonry building, it's got 16-inch [40 cm] thick walls, estimated weight around 600 tons … it's going to be a difficult challenge to move," said John, sizing up Rose's brick behemoth. Its bricks were soft and the cement holding them together was crumbling. Three towering chimneys ran up inside the structure and the building's ground plan was L-shaped. All these quirks made the building extremely hard to balance and move.

HOW TO MOVE A MANSION

The key to successfully relocating any building, whether it's a mansion, barn or lighthouse, is to keep the structure as level and stable as possible as it moves. "You have to fool the building into thinking it's not being moved," says Joe Jakubik, of International Chimney, who has project managed many ambitious moves. "In other words, you keep the bottom of the building level … supported at all times."

To achieve this, the first step in any move is to build a temporary foundation underneath the structure. Movers cut large holes in the basement walls of the building, just underneath the ground floor. Then they thread long steel beams through the holes. These stretch from wall to wall, across both the length and width of the building, crisscrossing underneath to form a sturdy, level rig.

Movers then position hydraulic lifting jacks underneath the rig. These are supported on towers of wooden blocks called cribs. Long hoses connect the jacks to a jacking machine. The machine controls the hydraulic pressure inside each jack and ensures they lift in unison. By increasing the pressure to the jacks, movers gradually raise the building off its old footings. The steel rig supports the weight of the walls and floors throughout, keeping them level.

While the building is supported on the cribs and jacks, the next step is to position sets of wheels, called

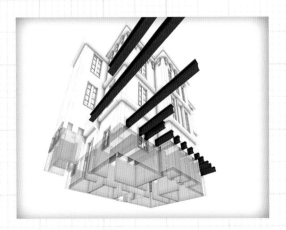

dollies, underneath the rig. The dollies usually have their own inbuilt power supply along with self-adjusting hydraulics. The hydraulics help compensate for bumps and dips in the ground, to keep the building level as it moves.

Once the building arrives at its new footings, movers line it up over its new foundation slab and gradually jack the building down to within a few feet of the ground. With the structure resting on crib towers, the movers build new walls up to the base of the building, leaving space around the beams so that they can slide them out. They then patch up the gaps, and allow the building to settle in its new home.

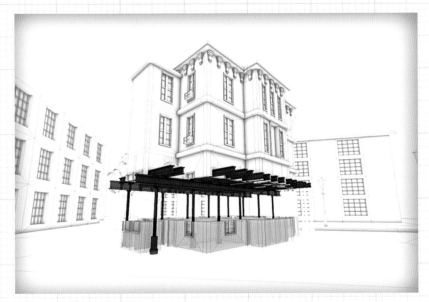

Above A latticework of steel beams is inserted underneath the ground floor of a mansion to support its walls during relocation.

Left Hydraulic lifting jacks push upward on the steel platform to raise the mansion off its foundations.

MOVING PIONEERS
THE INCREDIBLE EICHLEAYS

While the contemporary feats explored in this book are undoubtedly daring and inspired, many of them pale into insignificance when compared to the trailblazing exploits of a carpenter from Pittsburgh, Pennsylvania, called John Eichleay Jr. Between 1900 and 1940, his company embarked on a series of relocation projects that were ingenious and well ahead of their time. He turned the art of house moving into a science, establishing many of the ideas and techniques used by structural movers today.

One of his earliest and most impressive achievements was the relocation of a mansion up a 168-foot (51 m) cliff face in Squirrel Hill, Pittsburgh. The three-story home belonged to wealthy coal operator Captain Samuel Brown. It sat at the base of a cliff in front of a noisy railroad line that was due to be expanded.

The 75-year-old house had been built by Captain Brown's father. Brown was horrified at the prospect of the beloved family home that he had been born and raised in being torn down to make way for the new tracks. Fortuitously, he owned a plot of land at the top of the cliff directly behind the house. So he called in John Eichleay Jr. to relocate his home to this new site, which would be much quieter and had panoramic views of the Monongahela River.

Moving the 1,200-ton brick structure up the sheer rock face proved a formidable challenge. Eichleay seconded a crew of 30 workers to the project. They began by drilling and blasting the cliff with dynamite to carve out a series of six level steps up to the top.

With a steel platform built around the base of the building, Eichleay's crew positioned 300 hand-operated screw jacks underneath. At the blow of a whistle, they worked their way around the jacks, wrenching each one round by half a turn to raise the structure by ¼ inch (6 mm). Once all 300 jacks had been turned, a second whistle blew and the crew raced around the jacks again to give them another turn.

The mansion crept up the cliff face at the stately pace of 7 inches (18 cm) an hour. As they went, the workers propped up the building on towers made from wooden blocks. Once they had ascended each step, they used rollers to push the structure back along the ledge toward the rock face.

It was extremely dangerous work. The weight of the building pressing down on the edge of each step could have caused the rock to crumble or slip at any moment, spelling disaster for the mansion and almost certainly killing members of Eichleay's crew. But step by giant step they recycled 20,000 blocks of wood and reached the top in just 100 days without any accidents or injuries.

Tragically, Captain Brown's house burnt down 10 years later. But the feat is regarded as one of the greatest building moves of all time. It cemented Eichleay's reputation for ingenuity and many equally inventive schemes followed.

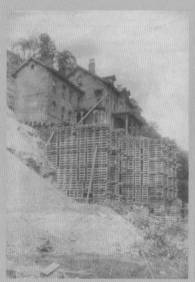

In 1915, American steel magnate Charles M. Schwab called in Eichleay to relocate his summer house. Schwab wanted to build a grander retreat on his Immergrün estate in Loretta, Pennsylvania. The best location was on the hilltop where his current house stood but Schwab was keen to retain the old building. So he hired Eichleay to move the house to another hilltop 1,500 feet (450 m) away.

The only problem for Eichleay was that a vineyard and several trees sat in the valley between the two hills. Schwab didn't want any of the bushes cut down or damaged, so Eichleay devised a daring plan to "leapfrog" the house over them. His crew erected wooden support trestles 36 feet (11 m) tall in between the vines and used a horse and tackle to drag the house over the treetops. The move was completed successfully in six weeks.

In 1918, Eichleay performed a unique operation on the home of Kodak founder George Eastman. Eastman wanted to extend the music room in his opulent mansion in Rochester, New York, to install a new pipe organ. The music room was situated in the center of the building, so to squeeze the organ inside, Eichleay sliced the mansion in half and slid one wing away from the other, opening up a 10-foot (3 m) gap at one end of the music room. With the organ installed in the gap, the crew sealed up the extension with matching brickwork. The move cost over $6 million – purportedly more than the mansion originally cost to build.

Eichleay's moves were so adept that he could shift entire city blocks while people carried on working inside the buildings. Staff in the hardware department at the

Joseph Woodwell Company in Pittsburgh continued to type at their desks while Eichleay's men shunted the towering eight-story brick structure 40 feet (12 m) away from the road to allow the street to be widened. During the one-and-a-half-day move, not a single window cracked or inkwell toppled over.

And in 1930, when Eichleay's crew was rolling the 12,000-ton Indiana Bell Telephone Building 100 feet (30 m) to a new site, they ensured that its 600-strong workforce were able to use the telephones inside during its eight-day journey. Even the toilets flushed!

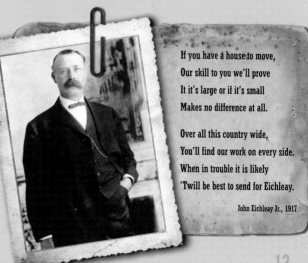

If you have a house to move,
Our skill to you we'll prove
It it's large or if it's small
Makes no difference at all.

Over all this country wide,
You'll find our work on every side.
When in trouble it is likely
'Twill be best to send for Eichleay.

John Eichleay Jr., 1917

THIS CALLS FOR PRECISION

Another factor that would make this job more testing for John's team was an extraordinary proviso added by Rose before she finally signed up for the move. To preserve as much of the estate's original charm as possible, Rose insisted that every historic feature on the farm – from its old outbuildings and ancient wooden fence posts to its paths, shrubs and plants – should be moved *with* the main farmhouse and realigned on the new plot in exactly the same orientation.

"I thought it was important that all the buildings be as they are today, because if they were not the way they are today then it's different and it's not the same homestead as it was on the location we are today," explained Rose prior to the move.

Work started in June 2006. John was anxious to complete the move before winter storms arrived and turned the site into a quagmire. The last thing he wanted was for the fragile brick building to be slipping and sliding around in the mud, so he had to work fast. Before his team could prepare the buildings for their journey, they had to work with surveyors to painstakingly map out the position and elevation of every building, paving stone and shrub to ensure that they could reinstate each feature at the new site in the correct setting. "Rose would like to be able to look out of her kitchen window and see the barn in the same place, and from the porch, see the milk house and the steam shed in the same place, and she wants the ... sunlight to come in the windows the same way," said structural engineer Kermit Christmann as he used GPS to survey and map the site.

Below Once the farmhouse had been lifted off its footings, movers supported its weight on towers of wooden blocks called cribs while they positioned wheels underneath.

THE MANSION
THAT JUMPED OVER A CHURCH

In June 2008, residents of upstate New York witnessed a truly bizarre sight: a historic mansion house literally leaping over a church. Hamilton Grange was originally built in 1802 for Alexander Hamilton, First Secretary of the Treasury of the United States. Over the years, the house had become run down and hemmed in by developments. Keen to restore the landmark to its former splendor, the National Park Service proposed relocating the structure to an open park nearby.

Moving the delicate wood-frame building weighing almost 300 tons would be a huge challenge. The front of the house was jammed into its plot by the porch of the church next door. Slicing the building in half to squeeze it out of the site was one option, but this would have compromised the building's integrity. Wolfe House Movers, based in Bernville, Pennsylvania, proposed a daring alternative. They would leapfrog the house over the church along a set of special steel tracks erected 35 feet (11 m) above the ground.

Work to lift the structure off its footings began in April 2009. The team installed a frame made from steel beams underneath the house and used hydraulic jacks to raise it off its foundations. As they lifted the building, they erected nine support towers made from wooden blocks underneath it.

When the Grange had reached its target height, the team built additional towers on the pavement and road, laying the steel tracks on top. They used more than 7,000 blocks of oak to build the sky-high slideway for the house. In a tense maneuver, hydraulic rams pushed rollers underneath the house to slide the building along the tracks. The towers held steady as the mansion soared over the church.

Once the Grange was hanging over the road, the team jacked the building down to street level, dismantling the towers as they inched it closer to the ground. The building's new location in St. Nicholas Park was just one block away. But the road had a steep gradient. Worried the fragile walls of the house might twist out of shape on the downhill run, the movers wrapped steel bands around the exterior walls to hold them tight. It took just two and a half hours to roll Hamilton Grange down the road to its luscious new setting.

Once the Grange had settled onto its new foundations, an ambitious scheme painstakingly restored its ornate features inside and out, preserving the landmark for future generations.

Above John Matyiko with his son Thomas on site

Before John's bulldozers and trucks arrived, botanist Lee-Anne Terry worked quickly to uproot Rose's garden. The original setting included many specimen shrubs and a walnut tree that was planted when the farmstead was originally built. Sadly, not all of them were candidates for moving. "Trees that have a fibrous root system are better candidates than this walnut," Lee-Anne explained to Rose.

Trees such as birch and elm whose roots spread out just below the surface of the ground can often be moved. Their roots can be supported on a steel frame and hauled to a new site by crane for replanting. But Rose's walnut tree had a deep tap-root system stretching up to 30 feet (9 m) into the ground. Its roots were too long, sensitive and old to remove intact. "The age of trees makes a huge difference," said Lee-Anne, examining the walnut tree. "This tree itself is probably close to 100 years old. Trees are a lot like people. When we are 40 we don't mind change, we can adapt to it. But when we are 70 or so, we don't adapt quite so easy. Typically trees that are about 70 years old or older do not adapt nearly as well as younger trees do." It took Lee-Anne's team of gardeners several days to label and dig out over 200 specimens. They were temporarily replanted on the edge of the farmstead while the buildings were shifted into position.

UNIQUE CHALLENGES

With the garden cleared, John's crew moved in to prepare the buildings for the move. Each structure posed a unique set of engineering challenges. The wooden walls and roof of the sprawling barn were old and rickety; the dual-layer brick walls of the milk house were separating; the farmstead was an unwieldy shape and extremely fragile. But the technology that house movers use to shore up and move such fragile structures is surprisingly basic.

Accounts and illustrations of 18th-century building moves show how wooden-framed structures were mounted on wooden wheels or tracks and pulled by animals to new locations. While powered dolly wheels and trucks have replaced animal power as the principal forms of traction, the materials, tools and techniques movers use to brace, balance and lift buildings have evolved little over the centuries. In fact, some of the relocation projects undertaken by early 19th-century pioneers were arguably far more complex and daring than moves in recent history (see pages 12–13).

Every structural mover deploys the same arsenal of timber battens, wooden crib blocks, steel beams and tensioning cables to brace and bolster the building for the road. The general rule is that the bigger and heavier the building, the more steel and wood shoring is needed. The house mover's greatest skill, however, is knowing precisely how to arrange the bracing and wheels around the structure to ensure that the load is evenly balanced, level and secure on the move. Amazingly, movers rarely use computer modeling to help them fathom all this out. They work

GABE MATYIKO: MANSION MOVER

My alarm goes off every morning at 5.30 a.m. I feed my yellow lab Sophie her breakfast, grab a bite to eat myself and head out the door to meet the guys at the shop at 7 a.m.

I give our crew equipment lists for each job, so when we show up the trucks are loaded up and ready to go to the site with the steel and tools we need for the day. The farther away from home the move job, the earlier we start to ensure that we get in a full day of work. In the summer it's pretty normal to work 10–12 hours a day, not including travel to and from the job. We try and make the most out of every hour of daylight we can get. Every so often the headlamps come out if we are a few hours short of finishing a job.

The majority of our moves that involve guiding a building down a road happen at night. The day before a night move we knock off no later than 5 p.m. to give the guys a chance to go home, eat, shower and get some rest. Everyone meets back at the shop around 10.30 p.m. so that we can be at the job early to make final preparations and checks before hitting the road. The average house move starts an hour either side of 1 a.m., and puts us on the new site between 4 a.m. and 6 a.m.

Most house-moving companies are small family-owned businesses. I am a third-generation house mover. It all started with my grandfather on my dad's side, Big John Matyiko (see page 107). Ever since I can remember, my father, Jerry Matyiko, took me on the job with him. I grew up doing this work every summer until I went to college.

After graduating, my wife and I took a year off to travel. When we returned to the U.S., a hurricane swept up the East Coast and caused devastating flooding all over the Chesapeake Bay area where I grew up. My father called me up and laid out the situation; hundreds of homes were in immediate need of being raised up. I'd always thought I wanted to do something other than house moving, I just hadn't figured out what that was yet. Needless to say I figured it out real quick and seven years later I can't imagine life any different!

Jerry and I share the workload of looking at new jobs, putting together estimates and managing the on-site move work. When the call volume is high, we can spend days on the road sizing up the jobs. Many of our larger projects require meetings with engineers, state highway officials and utility companies to get approvals for road moves.

For me, the hardest part of the job is how much time you have to put into it. The phone calls, on-site meetings, performing the work, night moves, working out of town – it adds up. Being away from my family as much as I am is hard for me and I have to thank my wife, Sarah, for being so understanding.

Jerry and I flew to Missouri to help out my cousin John Matyiko with Rose Heet's move. When moving a complex building the size of her farmstead, you can never have too many trained eyes on hand to monitor all the variables that go into a job like this.

The best way I can describe move day is by the relief that comes over me when the house is finally off the road at its destination. Even with all the planning and coordination that goes into a house move, the feeling you get in the pit of your stomach is unmistakable. There is no stopping, no backup plan, no detour. Once the house rolls out onto the road the only acceptable result is to make it to the new site. If an issue arises, you have to deal with it on the spot, with the clock constantly ticking.

House moving is hard work. You have to like working outdoors and not mind getting dirty. Being able to think outside of the box and having a "can do" attitude helps you get through the tough times. You have to be able to be a little hardheaded and stubborn, and like my father always says, "You've got to be flexible." Last but not least, it helps if you love what you're doing!

Hands down, the best part of the job is the sense of accomplishment that you get every day. There aren't many jobs out there where at the end of the day you can step back and see more literally the fruits of your labor. While you're working you'll hear people say things like, "That house was over there just this morning, now it's on the other side of the road!" or, "Hey, those guys jacked that house up 12 feet (3.7 m) in the air!" and it makes you smile. It sounds silly, but when you jump in the truck to head home after a long day and look back at the job, all the digging, block totting, driving, all that work is there in the form of a house in a different place than where you found it. That's house moving.

Above Braced and bolstered, Rose's machine shop trundles across the farm to its new location.

off gut instinct rather than reams of calculations. Years of experience listening to the creaks and groans of structures on the move seem to have taught these seasoned craftspeople how to devise low-tech, but infallible, forms of wrapping and strapping buildings for the road.

John's team built a network of wooden cross braces inside the chambers of the barn to hold its walls rigid. "Barns are often much tougher to move than houses," said John, inspecting the barn's dilapidated walls. "There's no actual flooring system to tie the building together." Rose's barn sprawled 37 feet (11 m) wide and 68 feet (21 m) long. Without bracing, the barn's walls would splay out as it moved. Similar struts held the milk house's interior layer of brickwork together, while 64 feet (20 m) of tensioned steel cables prevented its exterior walls from falling outward.

To keep the L-shaped farmhouse level as they lifted and moved it, John's team installed a huge steel support rig underneath the structure, made from three layers of steel beams up to 72 feet (22 m) long. Cutting holes for the beams through the building's 18-inch (45 cm) thick stone foundation walls proved arduous work. They had to remove as little masonry as possible to accommodate the beams, or the walls could have lost their strength and given way. It took John's crew of six workers over a week of laborious drilling to cut the 80 holes they needed to thread the massive beams right through the farmhouse. They packed the beams tight to the stonework with wooden wedges, sealed every gap and crack with cement and positioned 17 hydraulic lifting jacks underneath the rig.

PACKING UP

While John's team worked underneath the house, Rose's family worked frantically inside, taking the pictures off the walls and packing up a few precious antiques. John assured Rose that the homestead's fixtures and furnishings — from beds and sofas to tables and kitchen cupboards — would be safe left inside. This is the beauty of moving your entire house — there's very little packing up to do!

LOCATION, LOCATION, LOCATION

When developers began building a housing estate at the bottom of the Hodge family's garden in Suffolk, England, the family went to admirable lengths to retain the beautiful views they enjoyed from their Elizabethan mansion.

Originally built in 1593 by wool baron Thomas Eden, Ballingdon Hall was a stately oak-framed mansion sitting in the heart of Sudbury, Suffolk. The views from its leaded bay windows were stunning, stretching over 4 miles (over 6 km) across the Stour Valley and taking in church spires and lush rolling hills.

Through the years, the hall had been redeveloped into a junk shop then a hotel. But its breathtaking views and gardens had survived unspoiled. They were ultimately what attracted its current owners, Angela and John Hodge, to buy Ballingdon Hall in 1959 and convert it back into a home. The three-story building has over 20 spacious rooms. These include nine bedrooms, a stately dining room and a drawing room.

In the early 1970s, developers began building a new estate just beyond the Hodges' boundary at the bottom of the garden. The

development's modern homes and industrial warehouses soon impinged on the views from the mansion's windows. The development decreased the resale value of the Hodges' home, but made the land it stood on more valuable. However, the Hodges were reluctant to leave their beloved estate and move elsewhere.

At about this time, an ambitious effort in Egypt to relocate the twin temples at Abu Simbel had just concluded. The project to dismantle, move and re-erect the temples away from the rising waters of the River Nile had taken four and a half years and garnered international attention. (See Chapter 4, page 60.)

Inspired by this, the Hodges decided to embark on their own monumental move. They would relocate Ballingdon Hall to a new setting in their grounds. The new site sat ½ mile (800 m) across the landscape and 60 feet (18 m) higher. This would take the mansion away from

the development and allow the family to reclaim their spectacular views.

It was a courageous and high-risk decision. Unlike in America, where there is a long history of structural moving and a significant workforce of experienced movers, Britain's buildings rarely hit the road. Nevertheless, the Hodges tracked down a London-based engineering company called Pynfords who rose to the challenge of transplanting their mansion across the countryside.

The move was hugely ambitious. The structure weighed around 200 tons and no building this big had ever been moved this far, intact, in Britain before. The Hodges temporarily moved out to allow engineers to move in and prepare the hall for its trek uphill. Workers dismantled the brick chimneys and braced the flimsy wood and plaster walls with iron trusses. They constructed a rigid iron platform around the base of the building and secured 26 sets of wheels underneath.

On 29 February 1972 – leap-year day – Ballingdon Hall began its ascent. Unfortunately, the heavy building did anything but leap. Workers had to attach two caterpillar tractors to the structure to get it moving. Even with the extra traction, the hall moved only a few inches at a time. What was originally envisaged as a seven-day operation soon turned into a 52-day endurance test. The spectacle attracted more than 50,000 visitors, some from overseas. Spectators paid donations to charity for a front-row seat.

After the mansion finally arrived on its new perch, it took almost six years of meticulous work to restore Ballingdon Hall to its former glory. John Hodge reflected at the time, "I do not think it is worth anyone thinking they will have their house moved up a hill and believing when they have finished they can sell it and get their money back. However, from the point of view of us getting what we wanted, it has been worthwhile." Angela Hodge added, "What we have done may have been seen as somewhat eccentric, but it seemed like a good idea at the time. It may have taken a long time but I am delighted with it now."

With the farmstead braced and bolstered, John's crew connected up the jacks to a huge pumping machine. As the jacks powered up one by one, everyone's eyes turned to the building's foundations. The crew was anxiously looking for a horizontal separation crack to appear around the foundation. This would indicate that all the building's walls were lifting off the footings together. It was vital that the team kept the farmhouse perfectly level as they lifted it up. If one side of the farmhouse lifted faster than the other, there was a risk that the building could lean over, causing plaster and brickwork to crack. Over the course of an afternoon, the jacks raised the building 7 feet (2 m) up into the air, away from its foundations. Rose watched the lift with trepidation. "I just want to make sure it stays level," she said. "Otherwise, I could see everything toppling on the inside. And that would not be a good thing!"

With the building raised, the team spent a week lugging a staggering 144 wheels by hand into place underneath the house for the move. It was backbreaking work. By early October, bulldozers had finished clearing a level dirt path down to the new site and the farm buildings were ready to roll.

THE DAY OF THE BIG MOVE

Move day dawned overcast. Everyone was keeping their fingers crossed that the rain would hold off. Rose had put on a small barbecue and invited friends and neighbors around to watch. As the spectators took their seats, Rose climbed onto the farmhouse veranda. It was an emotional moment as she smashed a bottle of champagne against the farmhouse to wish it bon voyage. "Here's to a new day for an old house, for many generations to come," she said as onlookers cheered.

Below Balanced on 144 wheels, Rose's farmhouse drives along the dirt road to its new home.

MAGNIFICENT MANSIONS

Raleigh, the state capital of North Carolina, is one of the oldest and fastest-growing cities on the United States' East Coast, a major base for the country's technology and biotech giants. Its population has almost doubled in the last decade and a wave of new homes and skyscrapers dominate the landscape. If a time traveler from two centuries ago walked into town, one of the few parts they would recognize is Blount Street Village.

This suburb of eight huge, ornate mansions was established in 1855 to house state officials. Immaculately painted in picture-postcard shades of blue, green, beige and white, these were some of the grandest 19th-century homesteads in America.

One of the largest buildings in the estate is the Merrimon-Wynne mansion, a two-story wood-frame house built in 1875 for local lawyer and judge Augustus Merrimon. But in 2008, the magnificent mansions came under threat. A new development of 500 homes threatened to turn this historic village into a modern suburb. To save the buildings from demolition the state decided to relocate the mansions to empty plots across town and sell them as private homes.

Local architect Steve Shuster paired up with Greensboro house hauler Mike Blake to help preserve these precious mansions. "The Merrimon-Wynne is actually the most important historic home in the Blount Street project," recounted Steve. "It's certainly one of the largest, has much of the richest history and gives us, like any community maintaining a connection to our past, a real sense of place."

Navigating the enormous mansions round Raleigh's busy streets tested the skills of Mike's team. Each mansion had a unique set of antique features that required special care and attention. "Use a little addition, a little subtraction, a little multiplication, a little calculus, a little trigonometry and a lot of guesswork," joked Mike as he tackled each mansion with aplomb.

The Merrimon-Wynne house had three heavy brick chimneys that made it a difficult building to balance and keep level. Mike's team had to lock the base of each chimney into place with steel support beams to prevent them from dislodging during the move.

The Hume house, home of state horticulturalist Harold Hume, had sprouted a new wing since first built. This created a weak link between the two halves of the building. To keep the mansion knitted together during the move, Mike's team wrapped it in almost 650 feet (200 m) of steel cabling. "We are going to wrap this baby up, strap her real good; hopefully we won't injure her in any way but get her to the new location in one piece," said Mike confidently.

Huge crowds gathered throughout the summer to watch the historic mansions trundle to their new homes. "Three hundred tons of history on wheels ready to roll," beamed Mike on the day of the Merrimon-Wynne house move. "I think it's wonderful that they are preserving the old houses and keeping them in the community and moving them to a place where they can still be enjoyed," enthused one spectator.

Squeezing the mammoth mansions through Raleigh's narrow streets kept Mike's team on their toes. The Merrimon-Wynne mansion was so heavy that Mike had to attach a bulldozer to help pull it off its old site down the curb onto the road. "I think it's going to crash," remarked a spectator watching the tug-of-war that followed. "We have seen this house sitting here for months. It is surreal watching it go across the street," said a local as the mansions hit the road running.

One by one, over the course of summer, the mansions rolled in stately fashion across town. The odd flat tire and tight turn stretched the team's patience at times, but all the buildings eventually reached their new footings intact. Moving and restoring the village cost the state over $2 million. But the magnificent mansions soon became the most desirable residences in town and were preserved for generations to come.

The engine on the front of the house roared into life. John positioned spotters around the farmhouse to look out for any cracks or loose masonry that might appear as the building trundled down the road. Their first task was to guide the wheels across narrow wooden gangways off the old foundation hole and onto the dirt path. The ramps splintered and creaked under the building's tremendous weight as it inched forward. "The sooner we get off the boards, the safer the job is and there'll be a lot more smiles on guys' faces," said John's uncle Jerry Matyiko, who had come down from Maryland with his son, Gabe Matyiko, to lend a hand. It took the team an hour to travel the first 50 feet (15 m), repeatedly checking that each wheel remained level as they went. As they inched off the footings onto the dirt road, John's team leapfrogged large steel plates underneath the rows of wheels to prevent them from sinking into the dirt.

With the farmhouse on the straight and narrow, Jerry cranked up the engine. "I'm going to get them sweating. We've got to get over there before it rains," he urged. The farmhouse picked up the pace. Onlookers who had lost interest in the snail-paced race and gravitated around the barbecue, suddenly rejoined the steady procession across the farm estate, snapping photos. The sight of the immense structure wheeling down the road defied all belief. It was nothing short of awe inspiring, and a huge accomplishment for John's crew.

With the constant threat of rain overhead, the team worked frantically through the day to motor the farmstead to the new site. Keeping in mind Rose's insistence that all the buildings must be perfectly aligned, the final stage of the move was the most critical. As they approached the ramp down to the new footings, John set up

Below Rose Heet (second from left) with the crew from Expert House Movers, Missouri

a line of flags to guide the building into place. They chained heavy earthmovers to the rear of the house to stop it running away downhill during the final push. As they neared the new footings, the team attached plumb bobs to the walls to line them up precisely with the edge of the footings. John counted down the inches as the building neared the drop zone: "Four … three … two … one!" As the crowds cheered the farmhouse into place, Rose grabbed John and hugged him, breathing a sigh of relief. "I think it's incredible. It's so good to see it here in one piece. I knew you guys would do it!" she said.

The team spent the next few weeks moving the barn, milk house, machine shop and outdoor toilet into careful alignment with the farmhouse at the new site. It took several months to secure all the buildings onto their new foundations and for plumbers and electricians to connect up the utilities. Over the following months, Rose renovated the rooms in the house, giving the family home a new lease of life.

The huge upheaval of moving the family farm paid off for Rose. "The house is a lot quieter in its new setting. We don't have to yell at each other outside to be heard above the noise from the road any more!" she says. "And when we look out the doors and windows of the house today, we can see pretty much what we used to see before the move," she adds, relieved that the Voellinger estate survives for future generations of the family to enjoy.

Below Rose Heet's farmstead, restored to its former splendor, in its new location

John Matyiko's team led the operation to rescue Rose Heet's historic farmstead.

Clockwise from top left
The farmhouse is lifted up off its foundations and mounted on the moving rig; the machine shop and milk house are braced for moving; the farmhouse is moved along a dirt road balanced on 144 wheels; friends and relatives gather to watch the move.

Huge Homes

GIANT HOUSE JIGSAW

Not all house moves go according to plan. When veterinary surgeon Kathryn Finlayson decided to relocate her dream home from the center of a bustling town to a more peaceful plot in the countryside, events conspired against her. A runaway crane and hurricane storms were just two factors that turned the project into a tumultuous experience.

MOVE STATISTICS	
Manse House, Nova Scotia	
Material	Wood
Height	25 ft. (7.5 m)
Length	36 ft. (11 m)
Width	33 ft. (10 m)
Distance moved	19 miles (30 km)

D r Kathryn Finlayson runs a busy veterinary practice in the East Coast Canadian fishing town of Pictou, Nova Scotia. She works long and sometimes unpredictable hours, so the daily 18-mile (29 km) commute from her home in New Glasgow to her practice was becoming a grind. She tried to sell her house in order to move closer to Pictou, but couldn't find a buyer. "I have to be very close to where I work, because of situations like emergencies," says Kathryn. "I actually had tried to sell this house for probably about a year, but nobody really was interested in buying it, one because it was quite large, two because it was right on the street, and three, I think, because of how close it was to all of the other homes."

Kathryn's wood-frame house was built over 100 years ago for the minister of a nearby church. Its interior is packed with historic features including ornate wooden floors, doors, stair banister and several period fireplaces. Over the years, the building of new homes around it and an increase in traffic on the road directly outside the front door had left Kathryn's old Manse House crammed and noisy. "There wasn't a lot of privacy any more," she explains. "I just really, really like this house and thought there's no way I'll ever be able to build one that would look just like it. So I thought, well, I will look at possibly moving it."

Kathryn took the plunge and purchased a spacious plot of land in the Pictou countryside. The site was peaceful, located just off a country lane, and overlooked rolling hills. To transplant her town house to the countryside, Kathryn called in experienced father-and-son house-moving team Phil and Jason Leil, based at Truro, Nova Scotia.

Above The Manse House at its original site prior to relocation in 2006

A BIG JOB

Phil and Jason had moved many structures in the area before. But moving this huge house out of New Glasgow through its narrow roads would be a challenge even for them. "You've got three different options when moving a building of this size," said Jason, sizing up the job. "The first option would have been, if we were out in the country or somewhere where we had a lot of space, we could move the whole building in one piece." Moving Kathryn's home intact would have been the best way to preserve its character and integrity. But the route to the new site was lined with obstacles, including a bridge and many overhanging utility cables. "If we didn't have to move a lot of wires, we didn't have to move a lot of poles … it would be a great option. We'd put it on our trailer and we'd go with it," explained Jason.

A second option for the crew would have been to completely dismantle the wooden structure, moving it plank by plank. This approach works well with large rural buildings such as wooden barns. Their shape and layout are relatively simple, and their walls are rarely plastered over. This allows easy access to the joints, bolts and screws that hold them together, for dismantling and reassembly. Not only would dismantling Kathryn's house have damaged its plaster walls: the large building would also have proved extremely complex to reassemble. So Phil and Jason opted to cut Kathryn's house into four, smaller pieces for the move across town.

Top left Workers remove the fascia from the front of the house along the horizontal cut line, dismantle the chimney and support the front-door porch roof with wooden props.

Top right Cranes at the front and rear of the house lift the first piece of the sectioned building down to the ground.

Center left and right Cranes slowly lift the first section of the ground floor off its foundations onto transporters for the move.

Bottom left Half the house remaining to be moved from the old site

HOW TO DISSECT A WOODEN HOUSE

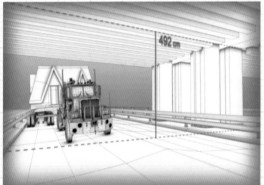

Cutting any building in half is a precarious operation. If movers make their incisions in the wrong place, leaving fragile or load-bearing areas of the structure unsupported, the walls could twist out of shape and the building would not fit back together properly.

The team dissecting the Manse House had to make two incisions through the building. The first one was vertical. The ideal scenario for the carpenters would have been to cut the building right down the middle. This would have minimized the width of each piece and made the move through the narrow streets easier. But this scheme would have jeopardized the house's integrity.

"If we were to cut through the middle it would leave floor joists from both sides sagging ... so the whole house would just ... tip in the middle," explained carpenter John Stokdijk.

Fortunately for John's crew, the house was built in two separate parts. The right, smaller wing of the house was a later addition to the main structure. So a

less risky strategy was to cut the building from top to bottom along the wing joint, bracing the exposed door and wall openings with wood for extra support. This maintained the house's rigidity, but made the move of the bigger, main section more difficult.

The second cut was horizontal. "More important than anything ... is the size of the piece, just to make sure that our height remains somewhat the same in the two pieces," explained John, wary that each section of the house had to squeeze under a bridge en route.

The horizontal cut line sat just under the second-floor windows. It preserved the windows that were old and fragile, but caused problems inside the house because it passed through the banister and upstairs door frames, which the team had to remove before they could dismantle the house. As they cut through the wall studs, the team bolted long steel beams along the length of each wall to prevent them from bending out of shape. The wood and steel supports ensured all four sections of the house retained their shape so that the house would fit back together at the new site.

Cutting a building in half isn't a task to be taken lightly. Although this plan would maintain most of the Manse House's characteristics and enable it to squeeze under bridges, it risked exposing the building's interior to the elements if it was not properly protected. Any rainwater seeping into the building could not only ruin its wooden features but also distort the shape of its frame.

"I was a little nervous about the floors and the banister and the fireplace," explains Kathryn. "I told them from the beginning that if they thought that the floors were going to get wrecked, or the fireplace mantel, or the banister, I might even not be interested in moving the house."

CUTTING UP A HOUSE

Kathryn moved out of the house in the summer of 2006, emptying it of furniture. Phil and Jason hired a crew of carpenters to begin the extreme surgery while the weather was good. Two huge cuts would split the building into four pieces. The vertical cut would cleanly slice off a whole wing. But the horizontal cut, running just below the second-floor windows, would pass directly through the doors, architraves, staircase and banister. So before the crew could begin dissecting the house, they had to remove these features to preserve them.

The carpenters devised a nifty coding system to ensure that they could reassemble every banister rod and door surround in its original position once the building had reached Pictou. They marked each piece of woodwork with a unique number and letter along with the area it came from so the pieces could be reinstated when the house was put back together.

The cut lines also passed through many water pipes. "Within the building we have pipelines from each radiator, we have pipelines from toilets, to bathtubs, sinks, washing machines, dishwashers," explained the move's plumber, Gordon Chapman. He too used a coding system to map out where all the pieces of disconnected pipe belonged. There were over 100 feet (30 m) of pipe work concealed inside the walls and floors, weaving in between the sections to be cut. It all needed mapping out and removing before they could pull the house apart into four separate loads.

CUTTING AND STRIPPING

While the carpenters stripped the inside, Jason dismantled the brick chimneys that snaked up through the two floors of the structure. Slicing the house in two horizontally would leave the top halves of the chimneys unsupported, without their foundation. The heavy towers could also lean as the structure moved on the road, putting pressure on the fragile wooden walls. Brick by brick, the chimneys came down. They would be rebuilt in their original positions at the new site, joining up with the fireplaces, which remained intact.

To create a clean horizontal cut through the walls on the second floor inside, the carpenters removed a strip of plasterwork to access the vertical studs inside the walls. The studs formed the skeleton of the house. "It'll make our cut a lot easier and we can see anything that might be in the way," explained carpenter

Above Carpenters removed the plasterwork on the lower walls of the upper floor to gain access to the vertical wall studs. They also removed the doors and door frames, and replaced the stair banister with a temporary wooden railing.

John Stokdijk. As the saw blades whirled into action, the inside of the house filled with clouds of sawdust and plaster powder. The noise was intense with carpenters stripping and slicing wood and plaster in every room. From the street outside, it sounded like the house was screaming as it was being torn apart. "It's like a big tornado hit it!" exclaimed Kathryn as she came by one afternoon to inspect progress on site. "This is a mess. This is very sad. They better fix it," she added, taken aback by the scale of the radical surgery inside.

STEEL SUPPORT

With every cut, the structure of the house became weakened. "Once we cut it all the walls will be hanging from the roof trusses," explained Jason. "That would allow all the studs to swing whatever way they feel like. They can bow in and they can bow out." To keep the walls rigid as they severed through the studs, the crew bolted lengths of steel across them. They skewered extra steel beams under the ground floor and below the second-floor windows. These would provide a sturdy frame underneath all four pieces of the building for cranes to latch onto and pull the building apart.

It took two days for the crew to sever through over 200 wall beams. By the time they had cut through the roof beams and secured all the supportive steelwork, it was heading into fall. With the house now exposed to the elements, it was vital that they finished the job quickly to avoid any rain wrecking the building. "There's no delay with this one," urged Jason. "We've gotta make sure we're ready to go and we've gotta make sure we can do it right away."

RUNAWAY CRANE

They needed two cranes to lift the main roof off the structure: one at the rear of the house, the second parked at the front, on the main road. But just as the driver tried to park the crane on the street in position, disaster struck. Its brakes failed.

Right The runaway crane smashed telephone poles and pulled down wires as it rolled down the hill. It came to a stop under a train bridge at the bottom of the road.

THE ULTIMATE OBSTACLE COURSE

Transporting a building from A to B is a huge logistical challenge. A keen DIY enthusiast might be tempted to hook up their own house to a truck to transplant it across town. But it's simply not viable given the huge hurdles even experienced structural movers face daily, trucking their superloads across town and country.

Depending on the house's size and weight, it can take weeks, sometimes months, of planning for movers to gain approval for a building to travel between two sites. Movers have to scope out a suitable route before they drive it with a house in tow, to ensure the building won't collide with bridges, clip overhanging tree branches or hit power and telephone cables slung across the road.

If a house is made of brick and particularly heavy, movers may have to work with structural engineers to find out whether roads and bridges en route are strong enough to take the strain. If the house is very wide, they have to obtain special permits to remove bus stops and road signs blocking the way. If the route passes through a busy town or city,

they may be forced to transport their loads at night to avoid rush-hour traffic. And if the house is traveling between states, movers will have to coordinate transport routes and timings with police departments and utility companies within each state. They also have to arrange special transport insurance in case the building or anything along the route becomes damaged.

The stakes for house movers are high. If the relocation projects are not meticulously planned out in advance, things can go dreadfully wrong. Take the case of Ricardo Gomez. In 1977, Ricardo contracted a house mover to relocate his house 5 miles (8 km) from Basin City to Othello in Washington, USA. Movers had to guide the house across a busy train line to reach the new site. Just as the house was crossing over the tracks, a freight train thundered down the line and crashed through the house.

"It was a like a bomb," recounts Ricardo, who witnessed his house being exploded into matchwood by the train that day. Unfortunately, Ricardo had also left his belongings inside the house for the move. They were now strewn across the tracks and pinned to the front of the train. "All the clothes, all the furniture ... everything," he recalls. Two brakemen were injured in the crash; both sustained cuts and bruises while one suffered cracked ribs.

Beware ... house moving is not for the amateur ... or the fainthearted!

SPLITTING UP

Every house move is an emotionally charged event for the home's owners. Perhaps the most extreme cases involve couples in the process of divorce. While most separating couples settle for mutually splitting up the CD collection and coming to a financial arrangement for the house, occasionally a husband or wife will want to literally split their bricks and mortar in two to hold onto a piece of their beloved home.

There have been several instances in America where divorcing couples have called in house movers to move the marital home from one plot to another site to allow one spouse to retain the house while the other keeps the land it stood on. In a case in Texas, a divorcing wife hired a move team to snatch a house off its plot while her soon-to-be-ex husband was out at work. He returned home to find his home had vanished!

In more extreme cases, movers are called in to literally saw a house in half so that the former partners get one half each. In 2008, an estranged husband in a village in southern Cambodia took matters into his own hands. Purportedly enraged that his wife would not look after him when he was unwell, he moved his possessions to one side of the home and cut the house in half. He then moved his slice of house to his parents' property, leaving his wife, after 40 years of marriage, to live in her own half, balanced precariously on stilts.

The crane rolled out of control down the street. Jason and Phil's crew made valiant efforts to try and bring it to a halt, throwing large chocks of wood underneath its tires to try and jam them. But the 35-ton machine kept on rolling.

The crane pulled down cables spanning the street and careered into telephone poles, tearing them in half. Traffic screeched to a halt to avoid colliding with the runaway crane as it gathered momentum down the hill. The only option for the driver to bring it to a halt was to jam it under a train bridge at the bottom of the road. It hurtled into the bridge with a crash. Luckily, no one was seriously hurt, but the incident shook everyone.

HURRICANE BERYL

The following day, the crew received more bad news. "The remnant of Hurricane Beryl is coming up on the east coast of Nova Scotia," sighed crew worker Sam. "It could be a tropical storm but it could … veer more to the east and we could just get heavy rain and a little bit of wind. But in Nova Scotia you never know what we're gonna have for weather." The team was now involved in a race against time. They had to dismantle, move and reassemble the house as quickly as possible, making it watertight before the storm hit.

With a replacement crane hooked to the roof with steel chains, they began to pull the first section away from the main structure. It was an anxious moment for Kathryn. "They just better damn well not drop it, or they're gonna build me a new house from scratch," she said from the street corner, watching the crane take the strain. With spotters positioned on the top floor to ensure the walls did not jam on the tall radiators as the roof section swung out, the crane drivers slowly inched it away from the ground floor. It was vital both crane arms moved at the same speed to avoid twisting the roof out of shape. Throughout the afternoon, in an extraordinary feat of coordination, the roof slowly swung over the top floor and crept down to the ground onto a specially built transport platform.

The spectacle was crowd stopping, resembling a giant doll's house being opened up for everyone to peer inside. Crowds of onlookers gathered on the street corner to watch the incredible sight. "That came apart good," said Phil, breathing a sigh of relief. "I hope it'll go back together as easy!"

ROAD RACE

With the storm advancing, the team was under tremendous pressure to complete the move. "We're lucky right now," said Jason. "There were no clouds in the sky when the sun went down. So far everything's going great for us. But it's Nova Scotia, it could change any minute." They made headway that afternoon. They positioned 112 wheels underneath the roof and hit the road at 1 a.m. The roof was the tallest section of the house. The pointed eaves over the windows narrowly avoided snagging on power lines overhanging the streets. It took the team 90 minutes to navigate 2 miles (about 3 km) through New Glasgow to the highway.

The wide road enabled them to pick up the pace, but a bridge looked as though it might scupper progress. The bridge was 16 feet 2 inches (492 cm) high, and the house was 1 inch (2.5 cm) too tall to squeeze under. The driver realized, however, that if they repositioned the truck wheels to ride either side of the crown in the middle of the road and lowered the truck's hydraulic suspension, they might gain a couple of inches' clearance. "It's gonna be tight," growled Sam as they inched forward with trepidation. Fortunately, they just managed to squeeze under. "We've never been that close before," grinned Phil with relief.

Left The Nova Scotia move team, including Phil Leil (second from left) and son Jason Leil (sixth from left)

HOMES TO GO

When the Daigle family from Vancouver Island, Canada, outgrew their modest three-bedroom bungalow, they rejected the idea of packing their bags and moving to a larger property. Reluctant to leave their spacious farm behind, they decided to have a roomier home literally shipped onto their plot.

They paid a visit to a unique yard in Victoria that sells secondhand wooden homes to go. "It's like a car lot, only with buildings," explained the yard's sales manager, Jim Connelly. The "housing lot" is one of several across British Columbia and Washington State set up and run by brothers Jeremy and Allan Nickel. The homes in the Nickels' yards are buildings that were due to be demolished. But Jeremy and Allan's company, which specializes in haulage, buys them for a dollar and moves them to the yards for resale. Buyers like the Daigles can explore the different secondhand homes for sale and walk away with a cottage for as little as $11,000. The Nickels then hook it up to a truck and deliver it to the new owners' doorstep.

By adopting these homes, families can not only pick up a bargain, but also do their bit for the environment. The United States produces over 100 million tons of building construction and demolition waste each year. That's enough to create a wall 30 feet (9 m) high around the coast of the country. Recycling houses in this way helps reduce this mound, saving on average 40–50 trees per home. And the period homes on sale have additional benefits.

"You get a better wood-frame construction than you could possibly build or buy nowadays," said Jim. "These homes have already stood the test of time, they've been against the elements for, in some cases, 100 years and they're still good to go."

With four kids, Darren and Tanya Daigle's eyes lit up when they stepped inside the Victorian show home. "It's almost double the square footage of our current house," exclaimed Darren, looking around. "It's got the style that we're looking for, lots of room for the children."

"Oh I'm really excited, I love this house," said Tanya.

Walking around the homes propped up on stilts in the Nickels' yard often proves nerve racking for potential buyers. "They're still a little bit taken aback by seeing

something that's up in the air on steel," said Jim. "I often see people having a reaction to it ... they walk very timidly through the building. They think that maybe it will fall over if they get in it."

Darren and Tanya bought the show home for just $25,000. The challenge of moving it 135 miles (217 km) north from the yard in Victoria to their farm in Courtenay rested on the shoulders of Jeremy and Allan Nickel. Jeremy and Allan have devised an ingenious way of transporting homes around the West Coast that avoids the heavy traffic and narrow roads. They ship them up and down the coastal waterways on large barges.

The sailing conditions around Vancouver for moving recycled homes from the yard or existing homes from one site to another are ideal. "It's not too often we get blindsided by a storm or wind," said Jeremy. "Generally speaking, what we advise people is that they can leave the furniture in the house. We'll ask them to take the pictures off the walls and take a few things off the shelves, especially if they're priceless ornaments. But we've had people who have just on purpose left half a glass of wine sitting on the counter, loaded a house in Victoria, sent it up island, it's arrived there and the half a glass of wine is still sitting on the counter in the glass."

While Jeremy loaded the Daigles' house onto the barge at docks in Vancouver Island, Allan was surveying a makeshift landing site on a beach in Courtenay, close to the family's farm. "I need to determine that there's enough water to bring the barge in and get it close enough so I can unload it off the barge," said Allan, measuring tide levels. The crew had to unload the Daigles' home at high tide to reach the road at the top of the shore. They laid down large metal ramps to form a sturdy bridge from the barge onto the road, and the Daigles' home trundled safely across, back onto dry land.

"Look at the size of it. What have we done!" exclaimed Darren as their new home turned into the driveway and tore through a "Welcome Home" banner. "The neighbors, they thought we were kind of crazy, but there it is." It took just a few weeks to lower the building onto its foundations and connect it to power and water, making it ready to move into.

The roof eventually pulled into the new site at 5 a.m. After just three hours' rest, the team was back at work preparing to move the first piece of ground floor so that it could be reunited with the roof before the storm struck. "We might just have to keep working right through tonight," sighed Sam at the prospect of another long day. "All of a sudden there's not enough time to sleep and lay down so you just have to keep going until it's done."

The first half of the ground floor hit the road just after midnight, reaching the new site at 4 a.m. As soon as the sun was up, the crew lifted the roof on top, painstakingly lining up all the wall studs of the top floor with those on the ground floor to ensure the house remained square. The carpenters worked through a third night stitching the first two sections of the house back together. Even though the roof was back on, the sides of both sections were still exposed. So the operation continued in earnest to move the second half of the house to Pictou.

THE GATHERING STORM

"The living room and dining room just went over the top of the roof!" joked a stunned onlooker as the crew used cranes to leapfrog the third section of the house onto the truck to move the next day. With the drivers under pressure to deliver both loads before the rain came, tensions ran high. "Get over. Get over! Pull over to the side!" screamed Sam to oncoming drivers as they raced the lounge to Pictou.

Below Inside, the main section of house is reattached to the upper floor. Outside, the floor of the side wing is exposed to the elements, waiting to be reunited with its walls and roof.

CAN WE TAKE THE POOL TOO?

When house hunters Wayne and Gloria McFarland from South Dakota, United States, wanted their dream home transported more than 70 miles (113 km) to a lakeside setting, one of the unique challenges faced by house mover Joshua Wendland was how to deliver it with both its indoor pools intact.

Wayne and Gloria had been searching for a new home for several years. They had kept in contact with Joshua, who runs Milbank House Movers Inc., a firm known across South Dakota for its relocation projects. Joshua regularly receives calls from people who want an old or disused house moved off their property.

When the owners of a large three-story Victorian house called in Joshua to haul it off their land to create space for a new development, he knew the historic home was perfect for Wayne and Gloria.

Built in 1905, the house had been well looked after, retaining its original woodwork and ornate fireplaces. It also had plenty of modern features, including a small swimming pool and a whirlpool. Wayne and Gloria bought the house and a vacant plot of land on Big

Stone Lake 70 miles (113 km) away for Joshua's team to move it to.

The house was one of the largest homes in Watertown, and too large to move along its roads intact. So Joshua's team cut it into three separate loads for the journey. The swimming pool posed an unusual problem. It was situated on the ground floor and its base sat on the house's concrete foundations with a wooden floor built up around it. There was a danger that once the team had lifted the house off its footings, the pool could fall through the floor without the support offered by its concrete underpinnings. So Joshua's crew bolstered its underside with wooden supports and used steel banding to tie it to the steel move rig built around the base of the house.

The three pieces of the McFarlands' new home traveled in convoy to the new site. "It went down the road like a parade of elephants," recounted Wayne. It took three days to haul the colossal convoy to Big Stone Lake. Once there, the house was put back together and its pools refilled, ready for Wayne and Gloria to move in.

ON THIN ICE

Trucking a heavy house across a frozen lake may seem like a crazy idea. But sometimes it's the only way that movers can deliver buildings to particularly isolated sites. In 2008, structural mover Kerry Neufeld, based in Saskatchewan, Canada, was called in to relocate a cottage from a shoreline plot on Emma Lake, near Warman, to a nearby golf course. A successful businessman working in the cellular-phone industry had bought the plot and wanted to build a multi-million-dollar mansion in the lakeside setting. However, he wanted to keep the pretty cottage. So he devised a plan to move the cottage to a golf course on the other side of the lake and offer it as a weekend retreat to potential new clients to help win business.

Driving the cottage around the lake to the new site was impossible. It was a tight squeeze along narrow roads lined with protected trees. The house was 42 feet (13 m) wide and would never fit. Sailing the cottage across the lake on a barge was also not an option. Dense trees also surrounded the lake. So getting a barge onto the water would have posed a huge challenge in itself. It seemed like there was no escape for the cottage, until Kerry suggested that the homeowner wait until the following March, when the weather was coldest, and then they would attempt to drive the house across Emma Lake, shore to shore, while the water was frozen. It was an audacious solution to the problem, but not without risks.

In the winter of 1899, movers in Salem, Connecticut, attempted to transport a house across Gardner Lake while it was frozen. They skidded the house, with its contents still inside, across the ice on a giant sled, pulled by oxen. By nightfall the house had only reached the halfway point. So they left it resting on the ice overnight. They returned the next day to find that the ice had cracked under the weight of the house, causing it to sink into the water. With no large machinery to rescue the house, it eventually sank to the lake bottom. Divers in the 1950s purportedly found toys and crockery inside its remains, while local anglers tell tales of haunting melodies occasionally emanating from the piano that was inside when the house went down.

Fortunately for Kerry Neufeld, technology has moved on a bit. Today, the traditional way to ensure that a frozen lake can support the heavy weight of a truck is to build a reinforced ice road on top of the existing layer of natural ice. This involves pumping thin sheets of water on top of the ice and leaving it to freeze overnight. Layer by layer, the thickness of ice increases. Ice-road builders regularly check the depth of the ice by drilling cores in its edge. The cores also allow them to monitor the ice's consistency. Man-made ice contains dirt and other impurities that make it weaker than naturally formed ice. So it often takes multiple floodings to ensure that the artificial ice road is robust enough to support the heavy weight of a house and truck.

Luckily for Kerry's crew, the winter that year had been particularly cold. Ice-core drilling revealed that the ice on Emma Lake had frozen naturally to a depth of about 4 feet (1.2 m), just thick enough to support the weight of the 80-ton house. So in March 2009, Kerry's team found themselves in a race against time to complete the move before the cold snap ended and the lake melted.

The old site was crammed with trees. So in order to haul the cottage to the ice road, Kerry's team had to cut it in half and transport the building in two loads. They severed the bedroom and hallway away from the main portion of the cottage, sealing the open ends with temporary wooden partitions. With a flat path across the ice cleared, they motored the two halves of the cottage across the lake one after the other to limit the weight on the ice road.

They monitored their speed carefully. If they traveled too slowly, there was a risk that the ice could crack around the load. But if they traveled too fast, a wave of water could form underneath the ice, rising up to crack the surface. So they maintained their speed at a steady 5 miles per hour (8 kph). Spotters walked alongside the trucks, listening carefully for any telltale cracking sounds.

After a tense two-hour journey across the ice, the two halves of the house were reunited in the new setting, ready for the owner to start schmoozing future clients with a round of golf in the spring.

Although they worked day and night for four days flat, the storm finally caught them. As the final piece of the jigsaw reached Pictou, high winds and driving rain made it too dangerous for the movers to crane the final section of roof into place. Rain poured into the house, pooling on the wooden floors. It also soaked the antique architraves that the carpenters had stacked up on the open top floor by mistake. They scrambled to seal the house with a tarpaulin. But high winds soon ripped it off and water still poured in, penetrating the ceilings and plasterwork.

The following morning, they lowered the final piece of the house into place. But there was a huge amount of restoration work to do before Kathryn could reclaim her home. It took many months of extensive renovation work and rebuilding to repair the water damage and make the house habitable again. Workers installed wooden ties to realign the warped wall studs, covering the walls with new boarding and plasterwork. Much of the original woodwork was ruined beyond repair and had to be replaced, while extensive restoration was needed to make the roof watertight.

Today the large town house looks a little out of place sitting in a lane of small country bungalows. But its views of rolling hills, clean air and tranquillity offer a welcome change of scene for Kathryn, who has since restored the Manse House into a beautiful family home.

Below Cranes lift the final piece of the jigsaw – the walls and roof of the upper wing – into place at the new site.

Extreme surgery saw Kathryn Finlayson's town house sliced up into pieces, moved and put back together like a giant jigsaw.

Clockwise from top left Cranes attached to the front and rear of Kathryn's home delicately lift the first quarter of the house down onto the ground, where it is positioned on a transportation rig; the second quarter of the house is lifted off the foundations onto the rig for transportation to the new site; cranes lower the walls and roof of the addition down to the ground, then leapfrog the living room and dining room over it onto the rig to move it to the new site before finally delivering the roof.

Colossal Churches

HOLY ROLLERS!
MOVING TRINITY CHURCH

Early one summer morning in June 2006, the residents of Lincoln County in rural Iowa woke up to witness a truly bizarre sight. Their local church was literally rolling down the road. Reaching a top speed of 8 miles per hour (13 kph), the pristine white building blazed a trail through the hilly landscape, resembling an ocean liner sailing through choppy seas.

MOVE STATISTICS	
Trinity Church	
Built	1913
Material	Wood
Weight	85 tons
Height	90 ft. (27.5 m)
Length	75 ft. (23 m)
Width	37 ft. (11 m)
Distance moved	11 miles (18 km)
Number of wheels	62

Built on a hill at the center of Lincoln County, Iowa, Trinity Lutheran Church had been this far-flung community's spiritual heart for over 100 years. Trinity's Lutheran congregation was established in 1881 by settlers arriving in the area from northern Germany. Dating from 1884, Trinity Church was one of the first structures to be built by the community. The parish remodeled it in 1901 to accommodate the growing number of worshippers. When a tornado completely destroyed the building in 1913, the congregation reconstructed their beloved church in just six months. The structure, standing on its grassy knoll, was maintained in immaculate condition. It was a cherished local landmark.

For generations, life revolved around worship at Trinity's altar. The rural sons and daughters of Iowa had their births and marriages blessed there, and their deaths duly honored. Records indicate that Trinity's congregation grew, at one time, to over 250 parishioners. With the area's declining rural population, however, Trinity's membership had dwindled to just 13 regular worshippers by 2006. No longer able to support it, but not prepared to see their dearly loved building demolished, the community raised the funds to have Trinity Church transported to a town across the prairies with more worshippers. It would be a wise but painful move.

FINAL SERVICE

"The one thing above all else we must remember is that the church moves on," reminded Trinity's minister, Bradley Ketchum, at the final mass held on 7 May 2006. "This church doesn't die today. It simply changes places." The service was packed out, drawing crowds from neighboring parishes. "As much as I have taken

care to preach in my sermons that this is only a building, it's a building that their fathers and mothers have in some cases spilt blood to build," remarked Ketchum after the emotional service.

The congregation's vision was to move Trinity 11 miles (18 km) to the town of Manning. The building would sit on a Heritage Park next to, rather appropriately, a large Hausbarn moved to the area, piece by piece, from Germany. Dan Peters from Manning volunteered to spearhead the relocation project. "It was a natural fit for the Heritage Park," recalls Dan. "The Park was established to preserve the legacies of the German settlement. Trinity and the Park fitted together well." In Manning, a thriving community would use Trinity for nondenominational Christian events including baptisms and weddings. The move would preserve a historic landmark for future generations.

Below Trinity Church in the 1930s

Bottom The remains of Trinity Church following a tornado strike in 1913

MOVING ON

The idea of moving a church to preserve it wasn't without precedent. There are many cases throughout history where abandoned or underused churches have been relocated to towns where parishioners had outgrown their current places of worship, or simply didn't have one. The cost of building a new church from scratch far outweighs the price of shipping in a "secondhand" chapel from a town close by. What would make the relocation of Trinity Church so unique, however, was its size and setting. Weighing over 85 tons and topped with a steeple 90 feet (27.5 m) tall, this immense structure's move across Iowa's undulating landscape would be a huge challenge.

Dan estimated that the cost of moving the church to Manning and constructing its new foundations would be around $100,000. The community proved extremely generous, raising the majority of the funds needed in just a few years. The next challenge for the project's organizers was to find a team of building movers brave enough to transport the structure across Iowa's rolling countryside. "We contacted six structural moving firms based around Iowa to tell them about the project and invited them to give estimates for the job," recalls Dan. Several declined to bid, worried the project was too big for them to handle.

The route to the new site was lined with precarious obstacles. The roads were made of loose gravel, extremely narrow and crowded with

RAILROAD RESCUE

In the 1960s, geological studies in the medieval city of Most, in the former Czechoslovakia, revealed that the area was sitting on top of 87 million tons of coal. Miners demolished almost every building in the historic city to access the coal. One of the few buildings to escape annihilation was its historic Gothic cathedral.

The Church of the Virgin Mary was built in the 14th century. Having survived a fire that destroyed a large part of the city in 1515, the building remained a fine example of Gothic architecture that many considered too valuable to destroy. To save the structure, but extract the coal, engineers devised an elaborate plan to use a special railroad track to relocate the cathedral to a new site ½ mile (800 m) away.

One of the most daunting challenges for the movers was the conservation of the cathedral's vaulted crypt that lay underneath the building. Preservationists demanded that to maintain the integrity of this historic but fragile feature, it should not be dismantled. So the movers had to devise a way to relocate the cathedral intact.

Workers built a reinforced "steel bathtub" around the base of the structure and installed a network of girders inside to give it extra rigidity. They positioned 50 trolleys underneath the building, running along four sets of train tracks. Hydraulic rams would push the trolleys along the tracks to move the cathedral to the new site.

The combined weight of the cathedral and steelwork totaled a massive 10,560 tons. It took the team 646 hours through the fall of 1975 to inch it 2,759 feet (841 m) onto its new foundations. The building traveled at an average speed of 1 inch (2.5 cm) per minute. Once the cathedral had been lowered onto its new basement, the steelwork was removed. The bell tower, which had been dismantled prior to the move, was rebuilt. The building reopened in 1988 and remains one of the few surviving relics of the city's past.

overhanging trees. There were sharp dips at every turn and a bridge to traverse. This was not a move for the fainthearted. Eventually, the organizers entrusted the move to mid-West structural mover Ron Holland. Working out of Forest City, Iowa, Ron had established a reputation for successfully guiding many sprawling structures, ranging from bungalows to barns, along the state's restrictive country roads. "We had moved other churches similar to it," said Ron. "The main thing was to look at the route ... the bridge looked OK and the trees could be trimmed." Ron's quote to relocate Trinity also came in favorably at around $25,000.

TRINITY'S TOWER

Ron's team of carpenters and mechanics arrived on site on 23 May to begin preparing Trinity for the road. One of their first challenges was deciding what to do about its iconic steeple. Movers often have to slice off or completely dismantle a tall bell tower before they can relocate a church.

Around the time of Trinity's relocation, a team of movers based in Sarasota, Florida, was tasked with relocating Crocker Church across the city to make way for an apartment block. The streets were lined with low-slung power lines, so to squeeze the building underneath, they had to sever off the steeple. This was no

HOW TO MOVE A CHURCH

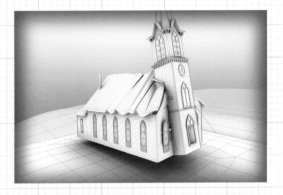

The key to ensuring Trinity Church survived its journey across the prairies was keeping the building as level as possible during its lift and move. It weighed 85 tons and this weight was unevenly distributed throughout the structure. The 90-foot (27 m) steeple alone weighed 12 tons. The steeple end of the church also housed the organ and a balcony. These features made that end of the church much heavier than the other, altar end. If the building tilted too much either as it was being raised or when it was on the move, there was a risk that the steeple could exert fatal pressure on the walls and roof, or become detached completely.

To lift the heavy building off its footings the crew built a rigid metal frame under its wooden floors from 14 steel beams measuring up to 82 feet (25 m) in length. They raised the structure up 6½ feet (2 m) off its basement with hydraulic lifting jacks. They had to apply extra pressure to the jacks under the heavy steeple end of the church to make sure that the entire structure remained level as it inched up toward the heavens.

At the side of the church Ron's team built a moving rig made up of 62 motorized wheels called dollies. They devised an inventive way to slide the church sideways onto this rig. They positioned sets of small rollers – resembling roller skates – underneath the church and chained winch trucks onto its side, one at the front and one at the rear. They had to ensure that both ends of the church rolled along guide beams onto the rig at exactly the same speed, so applied extra winching power to the heavy steeple end of the structure to guarantee a smooth, synchronized rollout.

To keep the building – and steeple – as level as possible during its trek across the countryside, the dolly wheels were fitted with self-adjusting hydraulics. These raised and lowered automatically to compensate for bumps and dips in the gravel road.

Together, these ingenious measures ensured Trinity reached Manning with its towering steeple still intact.

Above The move team had to keep Trinity Church level as they lifted and moved it to avoid its heavy bell tower from leaning and crushing the main structure.

Left The interior of the steeple was inherently well braced, so movers decided against dismantling it for the move.

easy task. Before they disconnected the tower from the main structure, the team had to build a wooden support rig around its base, thread straps underneath and attach them to a crane to take its weight. As the straps tensioned, they disconnected the steeple. They had to be on their guard against gusts of wind to ensure the fragile tower wasn't buffeted around as they lowered it to the ground. The chapel and steeple were eventually safely reunited at the new site a few days after the operation.

Ron Holland's crew carried out a full inspection of Trinity's tower. They were impressed with its solid construction, reinforced by its builders presumably to prevent it from suffering the same fate as its predecessor and being wiped out by a tornado. They opted to leave the steeple intact for the move, not wanting to risk compromising its integrity by having to slice it off and reattach it. Fortunately, out in the Iowa countryside there were very few utility lines across the roads that Ron's team would need to disconnect as they moved. The flipside of this decision was that the load would be unevenly balanced on the road. So they would need to design a clever wheel layout underneath the steeple end of the church, using additional wheels to spread the tower's 12 tons across a large road area to prevent the load from sinking into the gravel.

Below Trinity's move team, including truck driver Art Schulz (second from left) and Ron Holland (second from right)

Above Trinity Church moves down the hill off the perch it had occupied for almost 100 years. The team used the truck's hydraulic suspension to keep the steeple level as the building descended the slope and turned the corner onto the road.

FIRM FOOTINGS

While Ron's team installed the moving rig underneath the church, Dan oversaw the construction of the new foundations in Manning. Trinity would sit on a hilltop at the entrance to the town. There was a risk that in a heavy deluge of rain, something quite common in the state, torrents of water could pour off its steep roof and erode both the church's footings and the hillside. To prevent Trinity from sliding off its new perch, contractors had to design special gullies around the bottom of the building to direct rainwater away from its base and down the hill safely. The new concrete foundations and basement took several weeks to build and harden.

TRINITY'S ANTHEM

While both Ron's team and Dan's crew got Trinity ready for the big day, Manning's local choir, the Manning Quasquicentennial Chorus, were busy preparing a unique welcoming ceremony for the church. Choir leader Helen Wiese worked with musician Daniel Pemberton to compose and perform a rousing anthem called "Here It Comes" to mark Trinity's arrival in Manning. It would be sung live by the choir as they marched Trinity into town.

"The song is not just a testament to the life that breathes within and from Trinity Church itself, but also to everyone who, when faced with a dream that bit too big, has disregarded the doubters surrounding them, looked convention in the face, told it to get lost and carried on regardless," explains Daniel. The choir assembled for several sessions to rehearse their unique recital ahead of the big day. "'Here It Comes' captured the essence of the church move perfectly," recalls Helen. "This was a once-in-a-lifetime experience for our choir."

By early June, Ron's team had lifted Trinity off its footings, loaded it onto wheels and hitched a truck to the front. The load now towered 110 feet (34 m) high. Inside, Dan's crew secured the altar with ropes to prevent it from toppling over during what promised to be a bumpy ride.

THE STEEPLECHASE

At 7 a.m. on 6 June, the moment of truth arrived. News of the big move had spread to neighboring towns. Hundreds of onlookers from across the state gathered that morning along the route to witness this unique event. Many farmers parked tractors on hilltops and clambered onto the cab tops for a good vantage point. Some lifted their children above the crowds in the tractor buckets. There was a sense of great excitement in the air.

Ron briefed his crew ahead of the unique journey. "We do have some hills – that's gonna give us some problems," he warned. Father Ketchum fired the starting gun, saying a short prayer. On the chant of "Amen," Ron's driver, Art Schulz, fired up the truck at the front of the church and the move began. The church inched down off its perch at a snail's pace. "I've got a bet on the steeple falling off!" exclaimed one excited onlooker.

Ron's first task was to steer the church down off its hill and round a tight corner. Art inched down the road with trepidation. He raised the hydraulic wheels underneath the front of the bell tower as high as they would go to keep the building as level as possible. If it leaned too far forward, there was a risk the steeple could tear away from the main structure. Fortunately, it held rock solid. At the bottom of the hill, Ron's team, wearing hard hats, scurried underneath the church to steer its wheels around the first corner. Art carefully inched the extra-wide load around the bend and lined up the building with the gravel road. Trinity left its historic perch behind and trundled off through the landscape, motoring at 4 miles per hour (6 kph).

From a distance away, where the people scurrying around the building became indistinguishably small, the sight of the pristine white building cutting through the lush green fields against the clear blue sky was totally surreal. It was as if the invisible hand of God had plunged down from the heavens and was pushing the building along the landscape, playing with a life-size set of toys. The scale and

THE CHURCH THEY CUT IN TWO

In the mid-1950s, Reverend Theophilos P. Theophilos of St. Haralambos Greek Orthodox Church in Canton, Ohio, faced a pressing concern: overcrowding in his church. St. Haralambos was far too small to accommodate its growing congregation.

With no room to extend the church on its crammed plot, the parish purchased for $22,500 a vacant 10-acre (4 ha) piece of land 4 miles (6 km) away to build a much larger place of worship. The new church needed to accommodate 500 worshippers, twice as many as their existing building. Initial quotes from architects for constructing a new church, however, soon dampened enthusiasm. With costs soaring as high as $750,000, parishioners felt that this exceeded what they could raise for the project.

While on a visit to Cleveland, Father Theophilos caught sight of a hospital building being relocated to a new site. Armed with the name of the structural moving company

involved, he devised an inspiring new plan for his church. He would relocate the existing church building to the roomy new plot, and build an extension on the back of the building.

When representatives from the Cleveland moving company came to inspect the church, however, they realized that the massive brick structure was too large to squeeze through Canton's narrow streets intact. They suggested that they slice it in two – like a loaf of bread – to move it across town. They also proposed that instead of attaching a cumbersome new annex to the rear of the ornate structure, they could insert a new section of building 48 feet (15 m) long in the middle. Built from matching brickwork, the new extension would sandwich seamlessly in between the two existing halves of the church, doubling its capacity.

The cost of the move and remodeling was estimated at $200,000 – less than half the cost of building a new church from scratch. The congregation voted overwhelmingly in favor of the move. After two months of dedicated fund-raising teas and luncheons, parishioners had enough money to get moving.

Workers began preparing the church for the move in April 1958. They carefully took out its delicate stained-glass windows to prevent them smashing during the move, and removed the slate roof tiles to lighten the top-heavy load. As they cut the building in half, they shored up the exposed ends with a temporary wooden wall to prevent rain from wrecking the furniture that remained inside for the move. They built a support rig underneath each section from 50 tons of steel. With its two bell towers and domes, the front half of the building was the heavier piece, weighing over 300 tons. Jacks lifted it up off its foundations onto 60 solid rubber wheels for the journey.

At 2.20 p.m. on 14 June 1958, 5,000 spectators lined Walnut Street to watch the move begin. But as bulldozers dragged the hulking brick mass over the pavement onto the road, the wheels sank into the soft tarmac. Workers scrambled to lay wooden planks down the street to protect its surface. It took until 10 p.m. to free the church, when the operation packed up for the night.

News of the spectacle traveled fast. On day two, 50,000 people turned out to watch the spectacle unfold. City officials, worried that the church might fall through the street or down a manhole, insisted that the movers line vulnerable parts of the road with steel plates to spread the weight. The local gas company trailed behind the building, drilling test holes into the road to ensure no lines beneath had sprung leaks. Policemen directed traffic at intersections while the movers hauled traffic signals out of the way to let the church pass. At one bend, a winch line used to pull the building round the corner snapped under the strain. The church rolled backwards 3 feet (1 m) before workers stopped its slide.

After a nail-biting procession lasting 18 days, the first half of the church finally arrived at the new site. The second half was much lighter so completed its journey in just eight days. With both halves of the church lowered successfully onto the foundation slab, builders began to plug the gap between them. The only problem they hit was that the bricks used to build the original structure back in 1919 were no longer manufactured. So to ensure that the new and old slices of the church matched perfectly, engineers sandblasted its exterior to clean it up.

With its roof and windows reinstated and 48 additional solid oak pews installed, the church reopened on 1 March 1959, comfortably seating the entire congregation. The joins in the church that was cut in two remain invisible to this day.

Above Trinity Church is escorted across the prairies. A tractor waits ahead of the convoy in case the move team needs extra pulling power up the hills en route.

spectacle of the event was unforgettable. Locals scrambled into their pickup trucks and onto quad bikes to try and leapfrog ahead of the church to catch another glimpse. Many rallied behind on foot and a steady procession built up as Trinity sailed across the landscape.

PULLING POWER

A few miles in, the building ground to a halt at the foot of a steep hill. "The gravel has a tendency to lose traction," explained one of Ron's team. "We'll slip because it's like you're driving on marbles. You don't get the traction as if you were on black-topped concrete." Ron made a quick phone call. Shortly afterward, a stream of tractors raced across the hills to the rescue. The crew chained one of the tractors to the front of the truck. "I don't even know what it weighs, but it's got to be heavy," said the tractor's driver. "Plenty of power, plenty of traction. We shouldn't have any trouble." The extra 450 horsepower slowly helped inch the church up the hill and it steamed on through the afternoon at a leisurely pace.

THE INSIDE OUT CHURCH

In the 1920s, the population of the tiny British village of Blackhall, County Durham, exploded as people flocked to the area seeking jobs in a newly established colliery. Temporary homes were erected to house the thriving community. But Blackhall's local churches soon found themselves bursting at the seams coping with their rapidly expanding congregations. The churches were just too small to accommodate all the parishioners.

In 1929, the curate in charge of Blackhall, Reverend John W. Heap, devised an ambitious plan to "import" a larger church to the area from a neighboring town to solve the problem. St. Paul's Church in Wellington Street, Stockton, 16 miles (25 km) away, had recently closed its doors to worshippers. Attendance there had dropped off as Stockton's population had expanded and moved away from the church. The brick building, dating from 1885, was big enough to comfortably seat Blackhall's worshippers. So Rev. Heap envisaged moving St. Paul's to the town to form the heart of a new parish.

He bought the church for just $500. The building was too large to move intact. So he employed a building firm

based in Newcastle to dismantle, move and reassemble it, brick by brick. It took two months for the builders to dismantle the church. Three trucks a day shuttled more than 500 tons of bricks from Stockton to Blackhall.

The task of rebuilding the church began in February 1930. Having sat next to Stockton railroad station for over 40 years, the exterior of the church was blackened with soot and grime. To ensure that the building looked pristine when it opened, but avoid having to pay hefty costs for the brickwork to be cleaned, Rev. Heap decided that the building should be rebuilt "inside out." The dirty sides of the bricks would face inward and be plastered over, while the clean edges faced outward.

It took less than a year to reassemble the church. On 30 November 1930, the Bishop of Durham consecrated the reconstructed building, rededicating it to St. Andrew. By the time St. Andrew's Church had been completed, its parishioners had raised more than enough funds to cover the $10,000 cost of rebuilding it. At the time, building an entirely new church of similar size from scratch would have cost twice that amount.

A team of spotters ran ahead to keep an eye out for obstacles. "Watch the windows, guys," yelled one of Ron's crew as a tree brushed against Trinity's stained-glass windows. As it approached utility lines, engineers in bucket trucks temporarily disconnected them, lowering the lines out of the way to allow Trinity to pass. Fortunately, the church rode just high enough off the ground to miss hitting several mailboxes sticking up on the end of driveways. "I've never seen something this big go across my house before," gasped an onlooker as the church drove past her house.

HERE IT COMES!

Just outside the town of Manning, a police motorcade joined the church to guide it up the highway into town. Captain Lance Evans, who led the convoy, was slightly overwhelmed by the scale of this particular wide load. "We don't normally do things this big! We've moved all sorts of things from windmills to boats to houses, to now … churches!" he remarked.

Not far from the new site, Helen's choir corralled in front of the lead truck and marched Trinity to its new site bellowing "Here It Comes." "On its perch for all to see, safe and sound for eternity …" they sang, as the building rolled onto its new foundation to cheering crowds.

At 4 p.m., Ron's team, the choir and hundreds of locals gathered around the church for a celebratory barbecue. "It was a gorgeous Iowa day," recalls Helen. "Even though we seemed to walk and walk and walk, we all felt it was an amazing experience and worth the effort."

"Always a relief to get 'em there!" joked Ron. The celebrations lasted late into the night. Trinity was saved and looked at home on its new perch.

Today, Trinity is proving a popular attraction in Manning. "We do up to 15 weddings a year," says Dan. "We mainly cater for people who live within a 50-mile (80 km) radius in neighboring towns. But we recently had an enquiry from a couple in London, England. They were born in Iowa and are thinking of returning for their wedding." The steady stream of celebrations covers the cost of Trinity's upkeep. Its epic pilgrimage across the prairies saved this iconic structure for generations to come.

"HERE IT COMES"

One hundred years our church has stood
Pristine temple, built of wood
Abandoned, we fear losing it
To safety we are moving it

Here It Comes, here it comes, here it comes …

John Deere is helping us steer
Around tight corners in a higher gear
The heavy load of the church tower
Just won't move without tractor power

Here It Comes, here it comes, here it comes …

Climbing hills and crossing fields
Into town on sixty wheels
On its perch for all to see
Safe and sound for eternity

Amen.

Above Manning's Quasquicentennial Chorus leads Trinity Church into town.

Opposite Trinity draws crowds at its new location in the town of Manning.

Trinity's congregation witnessed the unique sight of their beloved church sailing across the prairies.

Clockwise from top right
Ron Holland's move crew guide Trinity past trees and posts along the narrow roads to the town of Manning; aerial photograph showing Trinity installed at its new site in Manning; inside Trinity Church; locals gather to watch Trinity's epic journey through the countryside; Trinity lifted off its old footings and balanced on the transport rig; Ron's team had to keep Trinity's tall steeple as vertical as possible as they transported the building across Iowa's undulating terrain.

Massive Monuments

RESCUING RAMESSES

When over 20 of Egypt's finest temples were threatened by flooding, 51 nations came to the rescue and mounted one of the most ambitious structural relocation projects in history. The engineers attempting to move the ancient temples of Abu Simbel found themselves locked in a race against time to literally move mountains.

MOVE STATISTICS	
The Great Temple, Abu Simbel	
Built	12th century BCE
Material	Sandstone
Weight	250,000 tons
Height	Facade: 108 ft. (33 m)
Length	184 ft. (56 m) into the mountain
Width	Facade: 115 ft. (35 m)
Distance moved	Raised: 213 ft. (65 m)

The colossal temples of Abu Simbel sit on the shores of the River Nile in Southern Egypt. They were built by the mighty Pharaoh Ramesses II – one of the longest-ruling pharaohs in Egyptian history. Ramesses' illustrious 67-year reign began in 1279 BCE. He was a great self-publicist, shamelessly proclaiming his own glory at every opportunity, building more extensively than any other pharaoh.

The Great Temple at Abu Simbel is an exceptional monument to his power, dominated by four statues of Ramesses himself, each towering 66 feet (20 m) high. A smaller temple for his queen, Nefertari, was built alongside. Both are impressive temples. But what's even more remarkable about them is that they were each carved into a solid mountain, just like Mount Rushmore. Unlike Mount Rushmore, though, hidden behind the imposing 108-foot (33 m) high facade of the Great Temple is a series of halls and chambers running deep into the heart of the mountain itself. Ornate statues and pillars line the hallways, with hieroglyphics adorning every wall of the twin temples. It took ancient engineers around 20 years to carve the temples painstakingly into the sandstone cliffs.

The Great Temple also possessed a special power. Ramesses' engineers carefully aligned its entrance so that twice a year the sun's rays would penetrate the monument to reach the inner sanctuary, lighting up the faces of statues of the gods carved inside. For over 3,000 years this extraordinary spectacle occurred on the same two dates every year – 22 February and 22 October – attracting thousands of visitors from around the world. But in the 1960s many feared this sunlight phenomenon would never be seen again. Abu Simbel was under threat from the very geographical feature that had enabled the ancient kingdom to flourish – the River Nile.

The Egyptian government needed to harness the power of the Nile by building the Aswan High Dam. This massive structure, almost 2½ miles (4 km) long, would control the river, which flooded the surrounding valley annually. In the plans, the floodwater would be contained in a reservoir – Lake Nasser – stretching for 300 miles (500 km) behind the dam wall. This would free up valuable farmland to supply Egypt's burgeoning population. The Aswan High Dam would also help to modernize the country's energy supply. Generators driven by the dam's water would provide enough electricity to power half of Egypt. But the consequences of the dam were far reaching. The resulting lake would flood the sites of more than 20 ancient monuments, including the twin temples at Abu Simbel.

INTERNATIONAL RESCUE

In 1960 the Egyptian government sent out an urgent plea to the United Nations for help. Experts from more than 50 countries gathered to debate the best way to save the temples. The most daunting challenge by far was how to deal with Abu Simbel's mountain setting. Various schemes were put forward to save the temples. "As you can imagine, when the project was launched, all over the world it fired the imaginations of the engineers, architects, even sometimes cranks," recalls Egyptologist Dr. Gaballa Ali Gaballa.

One idea was simply to allow Abu Simbel to flood and transform it into a colossal aquarium. Observation galleries would be built above the monument to allow spectators to peer down at the submerged temples. Special elevators would shuttle visitors underwater for a closer view. The biggest problem with this plan was that the temples are carved in sandstone, which is very porous. Over time, water would erode the stone, causing the temples to crumble. Another proposal was to keep the Nile at bay by building a giant permanent dam in front of the temples. The dam would have an elaborate drainage system to draw water away from the monument to keep it dry. However, the costs of constantly running the pumps would be huge. And the risks posed by the possibility of the drainage system failing were simply too great.

Engineers realized that the only way to save the great temples was to move them. The teams calculated that the flooding of the dam would raise water levels around the temples by around 200 feet (60 m). So to guarantee their safety, the temples would need to be raised up above the water level by at least 213 feet (65 m), and moved 655 feet (200 m) inland, clear of the shoreline.

Below The facade of the Temple of Queen Nefertari at Abu Simbel

THE UNDERWATER TEMPLE

The last temples on the rescue list sat on the island of Philae. These relics were already partially submerged when the relocation work began. Philae was once the center of worship for the goddess Isis and the focus of ancient pilgrimages. But by the late 1960s, it was in danger of being lost for ever.

Engineers built a temporary dam encircling most of the main buildings and pumped out the water. This allowed them to dismantle the temples on dry land ready to move them to a nearby island for rebuilding. But some of the island's sacred structures sat submerged, outside of the dam's reach. These included the Roman gateway to the island, the Diocletian Gates. Rescuing these from the bottom of the water required a different strategy.

A team of divers from the British Royal Navy was called in to help the salvage mission. "The first thought I had was how the heck am I going to get rid of the mud?" recalls team leader Lieutenant-Commander Ed Thompson about his first dive to assess the situation. Ed's team used a high-pressure hose to break up the encrusted clay, sucking the debris away with an underwater vacuum cleaner. Then they used hammers and chisels to take the structure apart. This was easier said than done. "Most things had to be done by touch because of the bad visibility in the water," recalls Ed. Each solid stone block weighed about half a ton. So the next problem was how to raise them to the surface. Their inventive solution was to attach canvas bags to each block. Inflated with air, the bags acted like balloons to float the stones up to the surface.

The low visibility and long hours made this rescue mission particularly dangerous. It took six months to raise the 450 blocks that made up the Diocletian Gates to the surface. Cranes lifted each block out of the water while barges shuttled them to the new site for rebuilding. The Royal Navy's mission was a complete success.

RAISING RAMESSES

It seemed like an impossible problem, particularly for the enormous Great Temple. How do you cut a temple that weighs around 250,000 tons free from its mountain setting and raise it to safety? The options were limited. The engineers first explored whether they could lift the entire temple to the new site in a single, intact piece.

One scheme they considered for doing this involved cutting around the entire monument to free it from the rock and then installing 650 hydraulic lifting jacks underneath. Lifting in unison, the jacks would raise the complete monument, weighing around 250,000 tons, one painstaking 1/32 inch (0.8 mm) at a time. Once the jacks were fully extended, engineers would replace them with concrete support pillars, reset the jacks, then lift again to raise the temple up to the next level. This process would be repeated over 200 times. But the risk was enormous. Nothing this heavy had ever been moved before. If any of the jacks failed, the precious temple could be damaged beyond repair. So this plan was rejected. "The difficulty of moving the temple en masse would have been hugely complex," explains engineer Richard Swift. "You're lifting something which weighs 250,000 tons. I'm sure that in 1964 that would have been a groundbreaker – literally!"

In another scheme, engineers proposed encasing the temples inside an immense concrete barge. As the water level rose, the barge would float the temple out of harm's way. But this idea was also abandoned from fears that a storm could damage the free-floating structure, or that the barge could rupture, causing the temples to sink.

Engineers realized that Abu Simbel was simply too heavy and delicate to move in a single, complete piece. They ultimately concluded that the only practical way to move the temples to dry ground was to cut up the monuments into small blocks, moving each one separately. At the time, this idea horrified archaeologists. Some went as far as to describe this option as pure butchery. "Cutting the monument was a very, very tough thing to accept," recalls Egyptologist Dr. Gaballa. "But you had to take this with that, the bitter with the sweet. This was the safest one."

Below Cranes lift and deposit each stone removed from Abu Simbel into a vast storage area.

HOW TO CUT UP A MOUNTAIN

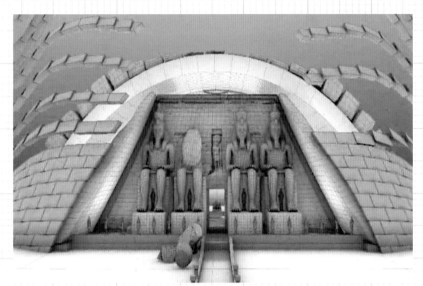

Left A concrete dome supports the weight of the mountain rock reinstated above the temple of Abu Simbel in its new location.

Below The inside and outside of the entire monument was cut into over 1,000 blocks.

Deciding where to make the incisions in the twin temples of Abu Simbel was a challenge. Cutting the temples into equal-sized square blocks would have made them easier to lift and reassemble, but could have disfigured the facade, leaving it looking like a chessboard. So engineers worked closely with archaeologists to draw up a careful cutting plan, tailoring each cut precisely to avoid slicing through a statue's nose or eye. Every single incision was meticulously plotted out before the cutting began.

It wasn't just the temple's facade that needed cutting. The rooms of the temple inside the mountain – their pillars, ceilings and hieroglyphic walls – also

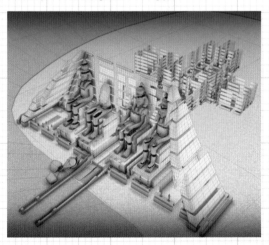

needed severing. Each room was decorated with historically significant scenes of Egyptian victories that could not be damaged. The archaeologists stipulated that no cuts were allowed across the faces of statues or on fragile decorated surfaces. But aesthetic worries had to be constantly balanced with technical concerns, such as the maximum size of block that the cranes could safely lift.

While the teams used light chainsaws to cut the backs of the blocks, every visible incision on the facade or inside the temple was done by hand using handsaws. Engineers drove very thin steel guide pins through slits in the rock to stake out the cutting lines. Passing from the front of the facade to the rear, they enabled workers slicing the backs of the blocks to keep track of the cut line at the front so that they knew where to meet up with it.

It was backbreaking work. The men slicing through the rooms' painted ceilings had to crouch in painful positions on top of scaffolding, manipulating the saws above their heads to dismantle the chambers. Each cut was made with the utmost precision and areas of brittle sandstone were injected with special resin to prevent them from crumbling. In total, over 5,000 individual cuts were made to slice the temples into 1,050 movable blocks, each weighing up to 30 tons. It took nine months of continuous sawing to dismantle the temples.

MOVING A MOUNTAIN

In 1964, an international team of 2,000 engineers and contractors from Sweden, Italy, France, Germany and Egypt began the monumental task of dismantling Abu Simbel. The scale of the operation, coordinated by UNESCO, was unprecedented. For young Swedish engineer Gösta Persson it would be the challenge of a lifetime. Gösta was a key member of the original team that went out to Egypt to pull off this amazing feat. "You could say we were trying to move a mountain, but you had to do it in pieces, we couldn't take the whole mountain in one piece," says Gösta, recalling the scale of the task they faced.

The engineers had only six months before the flooding Nile would be lapping at Ramesses' feet. To buy themselves more time, they built a temporary cofferdam in front of the monument to hold back the water. But it would give them only an extra 13 months before water would start pouring over the top.

Before they could begin to dismantle the underground temples, the engineers had to remove the 330,000 tons of mountain rock resting on top. They considered using explosives to do this quickly, but feared that the shock waves could damage the temples below. So they had to slice the mountain away using steel wire and chainsaws. It took 500 men seven months to cut off the entire mountaintop. Once the rock had been removed, they could begin the precarious task of dissecting the temples themselves into over 1,000 blocks.

REPLICA RESCUE

The heaviest, most precious blocks to be rescued at Abu Simbel were the heads of Ramesses. Each weighed 30 tons and was carved almost 66 feet (20 m) up into the mountain. In 2007 a team of modern-day movers, archaeologists and engineers attempted to investigate some of the hurdles the 1960s team had faced by using tools and techniques of the time to dismantle a replica statue.

Below, left to right
Stonemasons complete carving the replica statue of Ramesses from sandstone. Jonny Anderson (left) and Gabe Matyiko (right) slice the face off the head using a long handsaw. A crane latches onto steel rods cemented into the top of the head to lift the face away. The team then attach an L-shaped counterweight to the rear of the face to lock it back into the head.

A team of Egyptian stonemasons carved a replica head of Ramesses from the biggest sandstone block a modern quarry could provide. Weighing 10 tons, it was a mere third of the original size. American father-and-son structural moving team Jerry and Gabe Matyiko from Expert House Movers, Maryland, teamed up with British stonemason Jonny Anderson and engineer Richard Swift in a quarry in Cairo to plan and execute the move of the replica Ramesses statue.

The first cut on the statue would slice Ramesses' face away from his head. Before they made the incision, the team glued protective bandages around the cut line. Engineers had used this technique in the 1960s to prevent the edges of the sandstone monument from crumbling when cut. To reduce the risk of creating unsightly scars, instead of chainsaws the movers in the 1960s had used handsaws up to 11 feet (3.5 m) long. This way, each visible cut was less than ⅓ inch (8 mm) thick. Jerry's team in Cairo did exactly the same.

The modern team soon found out what an incredibly tough challenge the original engineers had faced when slicing up Abu Simbel. Sandstone is surprisingly hard, and it continually blunted their handsaws. The saws needed constant resharpening to ensure that they made a clean cut. And the team struggled in the searing desert heat.

After the first six hours of sawing, the team hit an unexpected problem. In a bid to help the saw make the cut, they had put oil on the blade. But not only did the oil stain the stone, it mixed with the fine stone powder to form tiny globules that clogged up the cut line. The oil actually made cutting much more difficult. They had to clear out the oily residue from inside the cut line with string, tugging it back and forth through the gap like giant dental floss, before they could carry on. Working in shifts with a group of Egyptian stonemasons, it took the team 12 hours to complete just a single cut – slicing Ramesses' face away from his head. It proved much harder than they had anticipated. "I wouldn't fancy doing two," joked Jonny.

TEMPLE ON THE TRACKS

ifty miles (80 km) upriver from Abu Simbel, a French engineering team battled to save the oldest monument on the rescue list – the Temple of Amada. The inner walls of its main sanctuary were adorned with important historical inscriptions and carefully painted plaster relief depicting key military campaigns. Slicing through this artwork to dismantle the temple wasn't an option. The sculpted plaster would flake off, ruining the temple.

Luckily this temple measured only 82 feet by 33 feet (25 m by 10 m), so engineers decided to move it intact, as a single unit. The team took great pains to protect the precious building, wrapping the temple's painted walls with cotton flock padding and bracing its crumbling corners with concrete supports. They wrapped the walls of the entire structure with high-tension steel cables to hold it together and threaded concrete beams underneath the floor to support the 900-ton monument for its lift and move. In an extraordinary operation, not unlike feats of the ancient Egyptians, hundreds of workers built a set of railroad

tracks to transport the temple across the uneven desert. Except instead of using manpower to pull the temple, they used hydraulic jacks to push it. The jacks were connected together to ensure that they exerted an equal pressure on the building and pushed in unison. This was crucial to prevent the temple from distorting or ripping apart.

The remote location of the Temple of Amada didn't help the operation. All supplies had to come from Aswan, which was three days away by water, or 137 miles (220 km) by desert tracks, where vehicles usually became bogged down in the sand. They could ship in only 500 feet (150 m) of track – a mere fraction of the 1½-mile (2.5 km) journey that lay ahead. So the team had to leapfrog the tracks, taking them up as soon as they were used and relaying them further up the route. They couldn't travel in a straight line. Instead the route had to zigzag around the uneven terrain of mounds and dunes. Moving at a rate of less than 82 feet (25 m) a day, the journey took three months to complete.

In the 1960s, the teams never stopped sawing. Five hundred men worked in shifts – day and night – for nine months. They made over 5,000 cuts – around six miles (10 km) in total length – to slice Abu Simbel into 1,050 movable blocks.

FACELIFT

For the Abu Simbel engineers, 10 October 1965 was a tense day. It was the day they lifted the first face away from the main structure. The stakes were high; 3,000 years of history hung in the balance. They feared that if they slung straps around the faces to lift them away with a crane, the straps could cut into the stone. So engineers developed an ingenious alternative: they drilled metal rods into each block. With the rods held firmly in place with cement, cranes latched onto them to lift each block to the ground. The rods acted like cocktail sticks picking up cheese. Engineers had to calculate carefully how far down into the blocks they drove the rods. If they didn't go far enough, then natural cracks in the rock could open up, and the bottom half of the face might break away from the top as it lifted. Engineers also had to mix ice into the cement to ensure it cured evenly and didn't crack in the searing desert heat.

Below Steel lifting rods protrude from the tops of the faces of the statues of Ramesses removed from the Great Temple at Abu Simbel.

When the team of modern-day movers replicated this part of the task, they encountered additional problems. The desert winds caused their face block to swing precariously in the air, even though they were lifting it just a few yards off the ground. The experiment offered a tiny glimpse of the challenges that the original teams had to contend with on a daily basis, working on an exposed mountain face.

Once engineers in the 1960s had successfully lowered Ramesses' face to the ground, they loaded it onto a truck. A support frame cushioned it on the flatbed while it made a ½-mile (800 m) journey to a special storage area. They had to drive extremely slowly to protect the fragile sandstone from vibrations. This procedure was painstakingly repeated over 1,000 times, for every block. In the central storage yard, the blocks were carefully cataloged and filed, like books in a library. Each block was given a unique identification code so that the engineers knew exactly where it had come from, and in which order the blocks would be reassembled once the new site had been prepared. The operation was like a jigsaw puzzle, only of epic proportions. So it was imperative that they knew exactly how to put all the pieces back together again.

Below Cranes remove the knees of the statues of Ramesses from the old site. The feet were the final pieces of the temples to be removed.

While work on the move at Abu Simbel progressed, more teams of engineers worked elsewhere along the Nile to rescue over 18 other sacred monuments. The design and setting of each temple made each move unique. Most temples were relocated to safe ground nearby. But some were transported further afield to museums in New York, Spain, Italy and the Netherlands.

THE GREAT DOME

The final pieces of Abu Simbel – the Pharaoh's enormous feet – were removed from the original site in April 1966. Four months later water surged over the top of the temporary cofferdam and flooded the site. But only half of the battle was over. Now the team had to reassemble the temples at their new elevation and rebuild the entire mountain around them.

To rebuild the mountain, the engineers needed to replicate the precise scientific properties that had enabled it to act like an arch holding

the weight of rock away from the ancient temples underneath. Simply stacking rock on top of the temples without any regimented structure could have caused them to fracture or collapse. The engineers' solution was to construct an enormous concrete dome over the reassembled temples, built from 24 interlocking concrete arches. A concrete structure of this height and span – stretching 89 feet (27 m) tall and 193 feet (59 m) wide – and bearing such enormous weight had never been built before. But the engineers' calculations paid off, and the dome carried the full weight of mountain rock piled on top, protecting the temples below. Block by block, they reinstated the front of the cliff face to ensure that the temples and their surroundings looked as if they had never been moved.

Above The statues of Ramesses under reconstruction at the new site. Engineers built a huge dome made from interlocking concrete arches over the inner temple to support the weight of mountain rock replaced on top.

The most challenging pieces of the puzzle to secure back into position were the faces of Ramesses. Cement was the most obvious solution but engineers were concerned that the bond might be too weak to hold the 30-ton weight firm. So they attached an L-shaped counterweight to the rear of each face with steel rods. The counterweight was shaped like a wedge and slotted tightly into a shaft cut into the Pharaoh's head, making it impossible for the face to fall away from its high mountain setting.

THE GREATEST TEST

It took 19 months to completely reassemble the temples at Abu Simbel. Engineers filled the 6 miles (10 km) of cuts between the temple blocks with mortar to cover up any signs of surgery. On 31 October 1968 – four and a half years after it had begun, and an astonishing 20 months ahead of schedule – the project was completed. Just as everyone had hoped, Abu Simbel was not only restored to its former glory, it looked as if the temples had never been moved. Engineer Gösta Persson is, quite rightly, proud of the team's achievement. "It's amazing … it's really fantastic. If you don't know about what has happened here, you can never imagine," he enthused, admiring the monument today. "It could very well have been here from the beginning. It's good, it's really good."

But the outstanding triumph of this greatest of all moves was that the relocation and realignment of the temples was so precise that today, twice a year, the sun's rays still strike the faces of the sacred statues inside the temple's inner sanctuary, just as they had always done for 3,000 years, and will do for centuries to come. "They were clever 3,000 years back when they carved this out of the rock," says Gösta. "We weren't so bad either when we moved them."

Right Once the pieces of the temples had been reassembled, engineers then had to remove the protective bandages glued around the cut lines and fill the gaps with mortar.

Overleaf The move was a triumph: it looked as if the temples had never been moved.

THE TEMPLE MOVED STONE BY STONE

The Temple of Kalabsha was an enormous edifice. Stretching 393 feet (120 m) long and 230 feet (70 m) wide, it was built between 30 BCE and 14 CE during the reign of the Roman Emperor Augustus. Unlike Abu Simbel, which was carved into a mountain, Kalabsha was a freestanding temple, constructed from 16,000 sandstone blocks. This form of temple should have been easier to move. But the temple's site was vulnerable to the seasonal flooding of the Nile, and sat submerged for nine months of the year.

With so many blocks to be moved, some weighing up to 30 tons, the German engineering team charged with its relocation soon realized that the three-month dry period wasn't long enough to dismantle the entire temple. So they decided to work *with* the water rather than against it. Rather than trying to beat the floods, they would make use of the changing water levels. They waited until the water was at its highest then encircled the temple with a flotilla of boats. Their plan was to dismantle the temple block by block, working down from the top, layer by layer, as the water level dropped.

Unlike Abu Simbel, where the teams raced against the rising river levels, at Kalabsha the dismantling process was

a race against the receding water, which fell by 20 inches (0.5 m) every day. They had just one summer to complete the task. There were so many snakes, scorpions and spiders on the land nearby that the workers had to live on boats instead of tents.

Five boat-mounted cranes carefully dismantled the temple, lifting the blocks onto reinforced barges that shuttled them 30 miles (50 km) to the new site. Disaster struck when the team was lifting one of the most precious of the temple's blocks, featuring an inscription of the fifth-century Nubian King Silko. The block slipped out of balance and fell onto the barge below. Luckily no one was killed. Stones already on the barge cushioned the block's fall, preventing it from sinking the boat.

The team finished dismantling the temple down to the foundations on 1 October. At the new site, workers used explosives to clear 423,000 cubic feet (12,000 m³) of rock and create a level base for the temple. They built a special harbor and road to deliver the blocks to the location. The project took two years to complete. Finally, in 1963, the 16,000 blocks were successfully reconstructed and the temple returned to its former glory.

In the 1960s, an international team raced to dismantle, move and rebuild the twin temples of Abu Simbel to prevent them from being flooded by the rising waters of the River Nile.

Clockwise from top left
The temples were carefully moved and rebuilt; the colossal statues of Ramesses II continue to stand guard outside today following relocation; the smaller Temple of Queen Nefertari to the right of the Great Temple was also relocated; sunlight still penetrates the inner sanctuary of the Great Temple on exactly the same two dates every year; Abu Simbel restored to its former glory.

Titanic Towns

THE FLOATING TOWN

Moving any building is a risky operation. But once in a while, house haulers take on the ultimate challenge of moving several buildings at once. Imagine the titanic struggle of moving an entire town ... across land ... or even water.

MOVE STATISTICS	
Float Home	
Material	Wood frame, concrete and polystyrene base
Weight	85 tons
Height	30 ft. (9 m)
Length	25 ft. (7.5 m)
Width	30 ft. (9 m)
Distance moved	170 miles (275 km)

Ken and Deanna Stratford lived in a regular suburban street in the town of Victoria, on Vancouver Island, Canada. After their children left home they were ready for a change of scene. Their dream was to live by the sea but waterfront houses around the popular Victoria Harbour were hard to come by. That was until Ken spotted an article in the paper about the plans of local developer Mark Lindholm.

Mark was about to transform an area of marina near the heart of Victoria. His idea was to build modern homes that would float on the water. A fleet of over 20 "float homes" would be moored in "streets" alongside a series of floating walkways to form a futuristic waterborne community. "You're building ordinary houses on extraordinary ground," says Mark, expounding the virtues of the floating-homes concept. "Everything is possible: we can put elevators in ... pretty much anything you do on land we can do on the water."

LIVING ON WATER

The idea of living on water isn't new. The Marsh Arabs of Southern Iraq built floating dwellings made of grass over 5,000 years ago. Houseboats with long histories can be found moored today in the waters of China, India, Thailand and Vietnam. In Europe, many of the barges and narrow boats that once worked the waterways have been converted into houseboats. And in the late 19th century primitive floating homes became common in the Vancouver, Seattle and Portland areas. "Float homes are part of the history of this area," recounts Mark. "When salmon fishing and logging were the major industries, workers often lived in houses out on the water. The guys who floated the timbers down the waterways would often use buoyant

cedar logs as floating rafts and build little houses on top ... My father lived in a floating home in Vancouver for many years."

Mark's floating homes would be firmly rooted in 21st-century design, fabricated from modern materials, fully plumbed, heated, powered and furnished with the latest fixtures and fittings. Ken and Deanna met up with Mark and were immediately sold on his concept. They took the plunge and signed up to buy a float home in his new marina. For Deanna the life aquatic sounded perfect. "We're hoping a float home will give us a lot of freedom. It certainly has everything that we would be looking for as far as lifestyle goes. You'll always have the light around you, you'll always be energized and interested in what's happening and what's going on."

THAT SINKING FEELING

Ken and Deanna had only one worry – sinking. How could they be sure that they wouldn't arrive home from work one day to find their house had sprung a leak and disappeared into the water? It was a reasonable concern. Large modern homes placed on floating footings had failed in the past. At the Venice Biennale art exhibition in 2009, artist Mike Bouchet displayed a full-size American family home set on top of floating pontoons. In the opening week of the exhibition disaster struck. One of the pontoons under the house gave way, causing the entire building to sink into the water.

To ensure that the Stratfords' new home was unsinkable, Mark's team designed an ingenious form of floating foundation, made from huge pieces of polystyrene coated in a tough, concrete shell. Mark was confident his designs for the float homes would be watertight. What he needed now was somewhere to construct them.

Below Workers seal large pieces of polystyrene foam inside a watertight concrete shell to form the base of the floating home. A wooden mold ensures the foundations set square.

The picturesque marina was no place for a building site. So Mark commissioned a marine fabrication company located over 170 miles (275 km) north of Victoria to construct the homes. Once built, the luxury floating mansions faced an epic sea journey to their final moorings in Victoria. With the move in mind, the construction team built two test homes for Mark's floating community on top of a special pontoon moored in the water. The pontoon had huge tanks below its base. When the tanks filled with water, the platform would sink, allowing the houses on top to float off into the water gently. At least, that was the idea.

It took eight months for the construction team to build the homes and prepare them for launching into the water. Float day was a tense time. The team released air from the tanks in the pontoon, flooding them with water one by one. Streams of bubbles erupted from underneath as the pontoon gradually sank. At one moment, the platform lurched precariously to one side. The houses tilted into the water and suddenly bobbed up off the pontoon, floating to the surface like corks. The fast-moving tidal waters around the launch site spun the homes dangerously close to each other. The team battled to control the buildings. They just prevented them from floating off down the river by lassoing them with ropes and securing them to moorings.

SETTING SAIL

The next stage of the operation was to tow the homes down to Victoria Harbour. The tides on Vancouver Island's east coast are among the strongest on the planet. Ebb and flow currents are channeled into powerful surges of water that can throw even the largest boat off course. When tugboat captain Kevin McGonigle arrived to

AMERICA'S GREATEST TOWN MOVE

"I believe there is iron under me, my bones feel rusty and chilly," exclaimed real-estate salesman Frank Hibbing one morning in December 1891, leaping out of a tent he had pitched in the middle of a freezing-cold pine forest in Minnesota. Hibbing, along with 30 other prospectors, had set up camp near Mountain Iron to dig exploratory pits in land he leased. There was speculation that the land contained deposits of iron ore. Hibbing had many doubters who had accused him of wasting time and money hunting for traces of the ore. But armed with little more than shovels and picks, they found rich seams in the first pit they dug. Hibbing had little idea at the time, but what he had stumbled across was one of the largest iron-ore deposits in the world.

Within the space of just a few years, Hibbing had established a small town in the area. He funded the construction of roads, homes, a bank, an electrical power plant and a hotel with his own money to draw people to Mountain Iron to help mine its riches. The town was named Hibbing in his honor. Hibbing expanded rapidly over the next two decades as iron from the mines fueled America's Industrial Revolution.

By 1915, around 20,000 people lived in the town. As Hibbing grew, many of the homes and businesses found themselves sitting on the edge of the mine itself, which was a noisy and dangerous experience. "With each and every blast, the rocks and shale had a most unpleasant way of coming down through one's roof, or giving one a sudden attack of heart failure, by falling in one's immediate neighborhood," recalled a reporter in the *St. Paul Dispatch* in 1918. So when further deposits of iron ore were discovered to be lying directly beneath the town itself, residents feared for their future.

To reassure Hibbing's inhabitants that they would not be forced to leave their beloved town to make way for the mine's expansion, the president of the Olivier Mining Company, W.J. Olcott, announced, "The Olivier Mining Company will not have to encroach on the improved district of Hibbing, with the possible exception of one or two lots,

for the next 15 years and possibly longer. The report the company is planning to move the village of Hibbing is without foundation." But with demand for iron rising sharply with the advent of the First World War, the residents of Hibbing's fears became justified.

To appease the townsfolk, rather than force them to leave their much-loved homes and municipal buildings behind and move into new dwellings elsewhere, the mining company funded an ambitious project to move the town 2 miles (3 km) south, off the iron-ore deposits, to join up with the neighboring town of Alice. It was a bold plan. In the scheme, around 200 buildings, including 185 homes, would be moved, street by street, down dirt roads to the new site. The relocation project, beginning in 1919, would cost around $18 million. A report at the time highlighted that "moving a town was a hazardous experiment without precedent."

The technology used for moving the buildings was rudimentary. Workers jacked up the homes off their foundations, resting them on timbers. To transport them to the new site, some laid rows of logs underneath the buildings. With horses strapped to the front, they simply rolled the houses down the road, leapfrogging the logs underneath to keep the building moving. Other structures were hitched to steam-powered crawlers, and driven to Alice on iron wheels. The movers relocated 14 buildings a week on average.

Not all the buildings survived the trip intact. One of the largest to take to the road was the Sellers Hotel. The three-story wood-frame building swerved off the road and toppled over. Along with the furniture and furnishings inside, it was totally wrecked. One worker was underneath the hotel when it collapsed. He was buried by the wreckage but rescued, and miraculously escaped without injuries.

The majority of the move was completed by 1921 and the town of Alice was renamed to Hibbing. The town and its inhabitants prospered in their new setting, while the mine itself expanded to become the biggest iron-ore mine in the world.

HOW TO BUILD A FLOATING HOUSE

The simplest way to make a house float is to build it on top of a hollow, air-filled structure, a bit like the hull of a boat. The only risk with this plan is that if the structure is holed it will fill up with water and sink. On a busy waterway, there's a constant risk that a ship could clip or collide with the base of a float home, causing it to spring a leak and sink.

To make the submerged foundations of their floating homes durable, but buoyant, Mark's team created a base made from giant pieces of polystyrene foam sealed in a thick layer of non-porous concrete. This concrete cast would prevent seawater from seeping into the structure if the base was damaged. The concrete was poured into molds arranged around the foam to ensure the base was square. Reinforcing metal bars were set into the concrete to give the base extra rigidity. Once set, the concrete molds were taken down.

To limit the buildup of algae on the sunken concrete footings, float-home builders often seal them with several coats of cold tar epoxy. These measures ensure that the base of the home will remain sturdy for at least 50 years.

Above the water level, the floating homes were built like any other wood-framed house. Ken and Deanna's had large windows to take in the views of the bay and passing water traffic and was finished on the outside with cladding.

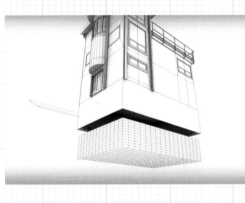

The concrete bases of floating homes are usually insulated and fitted with heating systems to keep them warm inside. The one thing that homeowners must watch carefully is how they arrange their belongings. The weight of bookcases, beds and furniture needs to be spread evenly across the floor space to keep the house level. A heavy piano in one corner of the room can result in the float home leaning nauseously to one side!

Left from top Some of the float homes were towed to Victoria using tugboats. Others were transported on a barge and lowered into the water by crane. The foundations of the float homes contain just enough polystyrene to enable the front door of the building to sit just above the water.

tow the float homes on their voyage south he knew that the tides would be a major challenge. He tied the two homes to his tug, one behind the other. But the flat-fronted concrete foundations were a cause for concern.

"The tides are going to really swing the houses in the water ... And I can't go too fast either," he said. "The homes aren't designed to be towed long distances. The square foundations will create waves in front as they're dragged through the water. If I go too fast the bow wave will get so big it might splash onto the front of the house and damage the wood inside ... We'll have to take it easy and watch the weather."

RISKY RAPIDS

The first day of the trip passed without incident. But on the second day, Kevin reached Dodd Narrow — one of the smallest and most dangerous straits in the area. Only 60 feet (18 m) wide and lined with rocky outcrops, it channeled the tides into vicious eddies and whirlpools. "It runs pretty good in there," said Kevin as he approached the infamous spot. "The tide runs hard. It can run up to six or seven knots." Kevin had only a 20-minute window to pass through the Narrow while the tides changed direction and the water was relatively calm.

WE'VE STRUCK GOLD

It's everyone's idea of a perfect dream – discovering your house is sitting on top of $5 billion worth of gold. That's exactly what happened to residents in the town of Malartic in Quebec, Canada. Quebec is home to some of the world's largest mineral deposits. Buried deep beneath the ground are vast seams of copper and silver, but the region's most valuable mineral is gold. A vast area of gold, called the Abitibi Gold Belt, stretches across the province. Hundreds of towns lie along the belt, exploiting these riches. Malartic is one of them, and one of Quebec's oldest "gold towns."

In 1926, exploratory drilling in the area led to the discovery of over 5 million ounces (140 million g) of gold beneath the Earth's surface. Settlers established a deep mine with over 15 miles (25 km) of tunnels to extract the gold. A few miles away from the deposits they built homes, schools and a church for the workforce.

The lure of gold turned Malartic into a boom town. But the prosperity didn't last for ever. By 1965, the gold had started to run out. Production eventually ground to a halt, leaving many unemployed.

With the gold depleted, people began to abandon the town. Malartic's future looked bleak until an extraordinary revelation offered a glimmer of hope. New exploratory drill holes near Malartic's church revealed undiscovered gold deposits buried underneath homes in the town.

In 2005, geologists from Canadian mining company Osisko moved in to investigate. They bored over 1,500 holes in roads, driveways and back gardens, hunting for gold. They analyzed over 250 miles (400 km) of core samples in laboratories as far away as Australia. The intensive drilling revealed that around 8 million ounces (225 million g) of untapped gold, worth over $5 billion, was buried directly beneath the southeastern part of the town itself.

The only problem was that the gold was sitting less than 1,640 feet (500 m) below the Earth's surface. If miners dug tunnels at such a shallow depth, the streets could cave in, destroying the homes. The only other way to excavate the gold was to dig it out from top to bottom and create an open-pit mine. But this plan would have obliterated Malartic completely.

The mining company decided that the only way to reach the gold without destroying the precious family homes on top was to move the southern part of the town to a new site over ½ mile (over 1 km) away. This way, they could safely turn the old area into Canada's largest open-pit mine, stretching over 1 mile (2 km) wide and ¼ mile (almost 500 m) deep.

The task of relocating over 150 homes was one of the largest moving operations ever mounted in Canada and cost over $60 million. In 2008, one of Quebec's leading heavy-haulage companies, Heneault & Gosselin Inc., was contracted to lift and truck the homes to the new site in less than 12 months to allow mining work to begin while the price of gold was high.

"Since it's a big project we've had to hire a whole bunch of new people," recounted Heneault & Gosselin's move coordinator Peter Tobin. Finding enough people in the area with the right skill set to move houses initially proved difficult.

"Since this is not something you can learn in school ... we basically had to start our own academy of house moving," said Peter.

He hired 15 new recruits to train in the basic skills of jacking and trucking houses. They practiced their house-moving skills on a mock-up building in the company's yard. Once trained, the team worked long hours on site, digging out, lifting and loading nine homes a week to stay on schedule.

Peter's crew had to move the houses at night, when there was less traffic on the roads, to avoid creating snarl-ups. The drivers carefully guided three houses at a time in a convoy, out of town, down the highway and across railroad tracks, delivering them to the new site before dawn. The homes were then lowered onto new basements, connected to electricity and plumbed in. With the gardens landscaped and roads tarmacked, the residents received new street addresses and new neighbors.

Peter's rookies performed the task admirably. The operation was completed on schedule, allowing Malartic's new gold rush to begin.

THE FASTEST TOWN IN THE WEST

American construction king David Cohen has grand plans for towns of the future. Cohen made his name working alongside his father building some of Denver's tallest skyscrapers. Having successfully conquered vertical building, David turned his attention to the horizontal. Today, there is more money to be made from building housing developments that stretch for miles over vast areas of land.

In 2006, David launched the first phase of an ambitious plan to build a community of over 200 homes in record time, reducing the time it takes to construct a four-bedroom house from 100 days to just 25. He believed that if you can build cars on a fast-moving production line, you should be able to do the same for houses.

On an empty stretch of land in Newbridge, Colorado, David erected the world's first factory for building full-size houses. His idea was to construct traditional wooden homes inside the huge warehouse. Once each house was complete, the team would move it on wheels to concrete foundations not far from the factory and set it down. With the factory set up at the heart of the new community, no home would have to move more than 1 mile (1.6 km). The factory would pump out a new house every week until the rows of empty streets outside were lined with homes. When the town was complete, the factory itself would be dismantled and moved to a new site where it could give birth to another new town.

Inside the factory, the construction team set up five workstations, one for each stage of the production. At station 1, cranes haul the large sections of walls, floors and roof into place like pieces of a giant flat-pack doll's house. Workers on gantries bolt the sections together, just like on a normal construction site, only indoors. The controlled environment means construction work can continue unabated during the winter months. At stations 2 and 3, plumbers and electricians install the pipe work and wiring. At station 4, decorators paint the walls and glaziers fit the windows. At the final station, they lay the floors, plumb the bathrooms and fit the kitchen.

Workers move the heavy homes from one station to the next on 10 air castors fixed underneath the base of the home. Each castor channels a powerful jet of air down toward the ground to create a thin, frictionless cushion of air under the house. "It's like a hoverpad," explains one of the workers. "You can actually push a whole house with your hand." Together the air castors can lift over 140 tons. They float the homes effortlessly from one stage of the assembly line to the next.

Just like on a traditional new housing estate, David's team displays a show home to attract new buyers. Customers pick the size and style of house they want from a brochure. There's also a wide range of finishings on offer for buyers to create their own bespoke bungalow. They even get a choice of positions on the street.

Once each home is complete, workers roll the house out of the factory door onto a special shuttle to deliver it to its concrete foundations. The shuttle is a platform made from steel beams. It sits on sets of hydraulic wheels and is steered by remote control. The hydraulics help compensate for bumps and dips in the road to keep the house level as it moves.

It takes a few hours to drive each town house to its street. Once the building reaches its footings, workers winch it onto the concrete foundations using steel cables. Taking delivery of a factory-built home proved a unique experience for the inaugural street's first occupants. "It is so weird to think that you can have your house made and brought out, and protected from all of this," remarked Sheila Rosales as she watched her pristine home being hauled into its garden in the midst of a snowstorm. "And it wasn't there two hours ago!"

With the first stage of the development complete and the novel concept proven, David's team has their sights set on creating more communities like this. "We developed this technology, but didn't know exactly what was going to happen until we were physically moving houses," explains David. "We're very excited about our future. House moving for us will become a daily activity. House moving is going to be taking on new dimensions."

Even with his 20 years of tugboat experience, Kevin knew it would be too risky to try to guide both float homes through Dodd Narrow on his own. So he radioed for a second tug to take one of the houses. "In the last 30 years I've been down here and retrieved four boats that have gone down," recounted Harvey, the second tug captain. "One guy read the tide wrong and he hit the up-tide and spun around and went under," he said nervously. With all eyes on the swirling currents, Kevin and Harvey steered the float homes through the straits with trepidation. Fortunately, both homes made it through safely.

A day later the maritime mansions finally pulled into Victoria Harbour. Mark's team inspected the homes. They hadn't traveled well. Bad weather and high waves had caused significant water damage. Mark needed to find a better way of moving the next batch of houses, which included Ken and Deanna's precious new home.

PLAN B

Mark and the construction team devised a new plan. Ken and Deanna's home would be erected alongside three other float homes on top of an enormous cargo barge. The team hoped that this vessel's oceangoing hull and high protective sides would make for a much quicker and safer transport method. The team built Ken and Deanna's home within the confines of the barge. The other three homes were assembled alongside and the barge set sail with the floating village resting high and dry on the deck. The spectacle of this shipment was truly bizarre; an entire street of homes sailing through the forested islands of the Pacific North West. Even with its heavy cargo the barge moved easily through the water and arrived in Victoria ahead of schedule.

Left Four float homes are transported to Victoria on the deck of a huge barge.

LIFT AND LAUNCH

Mark's next challenge was to get his homes off the ship and into the water. His plan was to dock the vessel at a nearby shipyard and use a giant industrial crane to hoist the buildings. Ken and Deanna came to the dockside to watch their house being swung into the water. The stakes were high. "This is their whole life savings they've put into these homes," remarked one of the workers, underlining the risks. "The last thing we want is a homeowner getting all upset because there was a scratch ... and who would blame them?"

The team threaded long nylon straps underneath the first of the four houses and attached them to the crane hooks. The crane team began lifting with trepidation. "I'd rather they were cautious now than after one's tumbled into the water," quipped Mark. The crane powered up. But the home didn't budge off the base.

The team paced the dockside looking for signs of movement. And to their horror, they then noticed that the entire barge was actually slowly rising up out of the water. The concrete foundations of the home, which had been poured directly onto the barge, were firmly stuck to the barge's surface. The crane was lifting both the house and the barge up out of the water! Ken and Deanna looked on nervously. Their house was positioned next door to the one being lifted. If the first house didn't separate from the barge smoothly, it could swing up and crash into the sides of the barge, or worse, it could collide with their new home.

The crane team had no choice but to slowly increase the lifting pressure and hope that the house didn't pop off too quickly. A crunching noise soon indicated that

Below Cranes lift the float homes off the barge and carefully lower them into the water.

something was happening. Then, with a satisfying crack, the concrete base of the home broke cleanly away from the barge. "We're free!" yelled one of the workers with relief. The house was now swinging in the air. "Get away from my house!" said Deanna as she watched the building sway dangerously close to her home, which was still sitting on the barge. Then in one smooth motion, the crane hoisted up the house and lowered it gently into the water.

It was a relief for everyone to see the first home bobbing around happily in the water. "Now that they've done the first one, then the next three, well, hey, they've broken ground!" quipped Deanna.

One by one, the homes sailed over the edge of the barge and into the water. The last house to be lifted off was Ken and Deanna's float home. Ken looked anxious as the crane began to lift. "It's kinda scary watching our whole future hanging from a couple of nylon straps!" remarked Ken. "Sounds like the story of our marriage!" replied Deanna with a smile. The house descended toward the water. "Now does it float? Three ... two ... one ... She's down ... all the way!" they exclaimed with joy.

TOWING THE HOUSES HOME

The four homes were roped to tugboats and guided into the marina. But the buildings weren't quite home and dry yet. "If you bring in the wrong house at the wrong time then it's in your way," said the marina manager as the tug captains battled to tow the huge homes past the water traffic and houses already moored in place. "If they start banging into each other then we're in trouble. One false move and you could take out a whole wall on one of these houses."

Above Cranes hook onto nylon straps threaded underneath the base of the float homes to lift them off the barge and launch them into the water.

Deanna could hardly bear to watch as her house pushed its way past the other float homes to reach its berth. "Gosh we're awfully close to that house there!" she gasped. "I'm like a nervous wreck! ... That is too close for comfort!" After a brush with a flagpole on the balcony of the neighboring float home, the team regained control of the Stratfords' house and slid it into position. They secured the home to its mooring with giant posts driven deep into the seabed. Ken and Deanna's new house was home at last!

"It's very much a relief to see the float homes get here without any significant damage," said Mark, watching the last home being secured in place. "All the float-home purchasers are very content with their homes so that's another relief ... It's a great sense of pride to see it finally coming together."

MOVING IN

Later that evening Ken and Deanna climbed aboard their new home. Its huge windows looked out on panoramic views of the bay, just as they'd hoped. All they had to do now was move in and get used to life on board a home that floats.

RISKY TOWN RESCUES

Once in a while, an entire city or town faces imminent danger and needs rescuing. In the 1850s, movers undertook an epic operation to raise the city of Chicago. Built on boggy marshland, its streets were often knee-deep in mud. To enable the city to prosper, engineers used screw jacks to raise its buildings 14 feet (4 m) higher. In a mammoth operation lasting 20 years, entire city blocks – some as long as 320 feet (98 m) and weighing over 35,000 tons – were lifted onto new foundations, intact, out of the quagmire.

A rescue of similar scale is currently under way in the Arctic mining town of Malmberget, Sweden. In the 19th century, the discovery of iron ore transformed the fortunes of this once-tiny community. Today, 15,000 people live and work in the town, which has a large sprawl of houses, shops and civic buildings.

Some of the richest deposits of ore sat 3,300 feet (1,000 m) directly beneath the town itself. A network of underground tunnels was dug to extract the iron out of the ground. But as miners removed each layer, a cavernous void grew under the town. By the late 1970s the walls of one of the voids had become so unstable that it steadily collapsed, creating a giant sinkhole on the outskirts of Malmberget.

With the hole getting bigger, and more sinkholes forming, the mining company took the drastic decision to move over 450 homes out of the town to safer ground 5 miles (8 km) away.

Moving each of the concrete homes costs around $120,000. But that's a third of the cost of building new homes this size. "This is a big job," said mover Andreas Martensson, surveying the streets. Together with his father, Magnus Martensson, he operates Tore Husflyttningar AB, one of the firms relocating the homes. "First of all we're going to move this street, and then we are going to move on further down the mountain towards the town and move more or less the whole of the town," said Andreas, undaunted by the task ahead.

The team endures formidable conditions moving each 200-ton house. Working in the Arctic Circle, during the winter they have just five hours of daylight a day and battle temperatures of just 5°F (-15°C). They must dig through several feet of snow and solid ice to free the houses' foundations. Driving homes 33 feet (10 m) wide through Malmberget's narrow streets is also a challenge. The project will last many years, but will ensure Malmberget can continue to prosper.

Electricity cables, water, gas and waste pipes were all hooked up through special flexible connections and the home functioned just like any other. Deanna found adjusting to life on the water even easier than she'd expected. "You just have to get used to the rhythms of the water – the tides, boats passing, that sort of thing. Sometimes you notice a lamp on a table swaying slightly as the house rocks gently in the water but that's all part of the charm. We've always loved being on boats and now we live on one complete with bedrooms, balconies and bathrooms!"

Mark now has 22 houses in his waterborne community, Westbay Marine Village. His marina is just about full. Elsewhere in the world, however, float-home communities are expanding well beyond the limits of small marinas and harbors. Floating structures provide a unique solution for populated areas with rising sea levels. In the Netherlands, where land for building is limited but water is abundant, float-home projects are becoming increasingly common. Floating towns with 40 or more houses have sprung up along its waterways, while large floating offices have been built in busy towns. The spectacle of floating buildings being towed along our rivers may way well become a common sight in years to come.

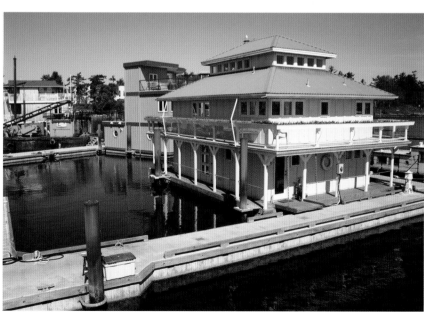

Left Metal posts driven into the seabed anchor the float homes into place alongside floating walkways.

Tugboat captains battled treacherous tides to ship futuristic floating homes down Vancouver Island's East Coast to create the waterborne community in Victoria Harbour, Westbay Marine Village.

Clockwise from top left
Four floating homes under construction on the deck of an enormous cargo barge; tugs guide the completed homes into place in Victoria Harbour; the homes are moored alongside floating walkways; the construction team "launch" the floating homes from a special pontoon; tugboats haul the homes from the construction site to Victoria Harbour.

Lofty Lighthouses

SAVING SANKATY HEAD LIGHTHOUSE

Perched on a cliff on the eastern shore of Nantucket, Massachusetts, Sankaty Head Lighthouse had been a guiding light for the island's seafarers for over 150 years. When storms left the beloved beacon perilously close to being swept away by the ocean, the islanders launched an epic rescue mission to haul it out of harm's way.

MOVE STATISTICS

Sankaty Head Lighthouse

Built	1850
Material	Masonry, iron and glass
Weight	450 tons
Height	75 ft. (23 m)
Distance moved	400 ft. (122 m)

Nantucket is a small island off America's Massachusetts coastline. During the 19th century, its small port was bustling with whaling ships, unloading their precious cargo of whale oil. Islanders built a series of lighthouses around Nantucket whose beams and foghorns could guide ships through even the thickest fog. Of all these lighthouses, Santaky Head Lighthouse, built in 1850, remains the most iconic. Towering over the village of Sconset on a bluff 200 feet (61 m) high, its red-and-white striped tower can be seen the length of the island. Today, Nantucket's whaling days may be over, but Sankaty Head Lighthouse is still a huge part of the island's identity. "It is a touchstone for the island," says summer resident Betsy Grubbs. "People see it when they come into the harbor, so it's really an icon for the whole island."

But, like so many parts of America's eastern seaboard, Nantucket is changing. Winter storms and Atlantic hurricanes batter its shores, ripping off huge chunks of coastline and threatening the wooden shingle-clad homes built close to the waterfront. Bob Felch, who lives in the village of Sconset, has noticed these dramatic changes. "The erosion at the bluff edge here at Sankaty Head has accelerated to an average of 3 feet [1 m] of loss a year. And what I have been telling everybody is that we are one storm away from being in trouble."

Sankaty Head Lighthouse is not the only building in danger. More than 10 luxury houses on the island have already been swept away. Islanders have spent hundreds of thousands of dollars on coastal armoring. But attempts to stem erosion with the likes of sandbag walls have had little effect. Nantucket's eastern coastline lacks solid bedrock and the loose sandy soil is no match for the mighty power of the Atlantic Ocean.

BRINK OF DISASTER

When built, Sankaty Head Lighthouse sat a comfortable 280 feet (85 m) away from the headland. But the cliff edge has gradually eroded away. Almost 16 feet (5 m) of Sankaty's cliff face collapsed during 1991 alone, in the year of the "Perfect Storm," when 60-mile-per-hour (100 kph) winds blasted the coastline for five consecutive nights. By 2005, the tower teetered around 80 feet (25 m) away from the bluff. Time was running out for Nantucket's "blazing star." The islanders faced a daunting decision. Did they surrender the lighthouse to the waves, or somehow try and salvage the building? For the islanders there was no debate. The lighthouse had to be rescued, so they launched a scheme to move the building away from the precipice, further inland to safety.

The islanders spent several years raising over $3 million to fund the ambitious rescue mission. Nantucket has some of the most expensive real estate in America, with a steady flow of millionaires flying in for the summer vacation. Through special fund-raising dinners and generous donations, by spring 2007 the Sconset Trust, the local group coordinating the project, had accrued enough resources to haul their beloved lighthouse out of harm's way.

MISSION IMPOSSIBLE

Moving the towering 75-foot (23 m) lighthouse off the edge of the bluff would be a precarious undertaking. The land around the lighthouse was extremely unstable. Turning the cliff top into a building site with heavy, thumping machinery could further compromise its stability and cause more chunks of the headland to collapse

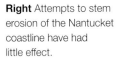

Right Attempts to stem erosion of the Nantucket coastline have had little effect.

into the sea. The islanders needed a team with a delicate touch. They called upon the services of two companies with unrivaled expertise in restoring and relocating tall structures. Rick Lohr's group of engineers at International Chimney in Buffalo, New York, had forged a formidable partnership with structural mover Jerry Matyiko of Expert House Movers, Maryland, in the past to rescue three of America's most iconic lighthouses, including the country's tallest, the Cape Hatteras Lighthouse in North Carolina (see page 104). The islanders entrusted this alliance with saving Sankaty Head Lighthouse.

"If Expert House Movers doesn't move this building, then Mother Nature will, and I don't think you will want that," said Jerry Matyiko, surveying the building

Above Jerry Matyiko of Expert House Movers, Maryland

while chomping on his trademark cigar. Jerry and Rick's crews had moved many structures successfully in their time together. Even though Sankaty Head Lighthouse was smaller than many of their previous conquests, the building's unstable footings and fragile brickwork – soft from decades of pummeling by storms – made this mission uniquely challenging.

In July 2007 the islanders crowded onto the headland to watch Sankaty's precious light being temporarily switched off. They clasped their hands together to form a chain around the lighthouse and give the building a ceremonial "hug." "It's the most important little place on this island," said teary resident Joan Porter, who had "hugged" the lighthouse every year for the last 77 years. "Oh Sankaty, we love you ..." sang the group, raising glasses to toast their favorite building on its way.

RACE AGAINST TIME

The next day, the team moved in. Rick and Jerry's crew had just two months to complete the move before the onslaught of the hurricane season. They staked out the cliff top, dividing it up into safe zones where they could spread out their equipment, so as not to overload the crumbling bluff. Their plan to move Sankaty involved mounting the lighthouse on steel guide rails and using hydraulic rams to push it steadily along the tracks 400 feet (122 m) inland, away from the headland.

Before the team could begin work on the move, they needed to shore up the building's weak brickwork. "This is typical of moisture getting behind the paint," said Jerry, inspecting the masonry and pulling a chunk of brick off the building with his hand. It was crumbling like feta cheese. "It freezes, and the face of the brick pops off and then you get more moisture in it and then it freezes and more pops off." Rainwater freezing in tiny fissures in the masonry had created a series of narrow cracks up the tower. There was a real risk that tiny vibrations created during the move could cause these cracks to widen, permanently damaging the lighthouse walls.

To reinforce the brickwork, the team built a support collar around its exterior. Wooden struts held in place with high-tension steel cables gripped the structure in a rigid embrace. "All right, we are going to test the tension on this," said Jerry, marching around the collar with a sledgehammer and his cigar. "If the cigar falls off it's not tight enough," he explained, perching the cigar on top of one of the wires and whacking the cable hard with the hammer. The cigar stayed put each time. "So you can see that each one of them is taut," concluded Jerry. With Sankaty firmly wrapped and strapped, the team could safely begin preparing it for the move.

FIRM FOOTINGS

Their first task was to build a temporary foundation underneath the lighthouse to support its base while it was pushed along the tracks. Moving a tall building with such a narrow base presented obvious problems. If the building leaned as they lifted or moved it, the tower could easily topple over. And such a concentrated load could destabilize the sandy ground as it traveled over it. So Jerry worked with International Chimney's Project Manager, Joe Jakubik, to erect a stable platform around the structure to spread out its 450-ton weight and ensure it remained level. "The existing foundation is essentially the size of a circle. What we are going to do is increase the size of the foundation," Joe explained, examining plans of a large square support frame he intended to build under the lighthouse. "We are taking this very concentrated load and spreading it out over a much larger frame."

Below A steel support frame erected around the base of the lighthouse spreads out its 450-ton weight over a larger area and keeps it level as it is pushed along rails inland.

HOW TO MOVE A LIGHTHOUSE

Moving a tall, skinny brick tower is a risky operation. Relocating Sankaty Head Lighthouse, perched on the edge of a collapsing cliff, was no exception. Most brick structures are loaded onto sets of wheels to move them from one site to another. But with unstable, bumpy sand beneath their feet, the movers of Sankaty Head Lighthouse needed a smoother, steadier ride to keep the tower from toppling over. So the team flattened a 400-foot (122 m) path in the sand stretching from the old foundations to the new site and laid two parallel tracks of steel rails along the roadway. These would act as guide rails for the lighthouse, which would roll on top.

Underneath the base of the lighthouse, the team constructed a temporary foundation made from three layers of interlocking steel beams. Hydraulic lifting jacks pushed upward on the steel rig to release the lighthouse from its foundations. A spiral staircase made from 62 cast-iron steps ran down the center of the lighthouse. The steps were threaded like beads along a steel column anchored firmly in the lighthouse foundations. To prevent the steps unwinding when they lifted the lighthouse off its footings, the team suspended the staircase from the building's upper floor with a web of tensioned steel cables.

The movers set small steel rollers between the rails and the steel rig to enable the building to roll smoothly along the tracks. Two push jacks positioned at the rear of the lighthouse thrust the building forward along the rails. Once the lighthouse had traversed along a length of track, the team leapfrogged the beams around to the front of the building, gradually extending the steel rollway to the new site.

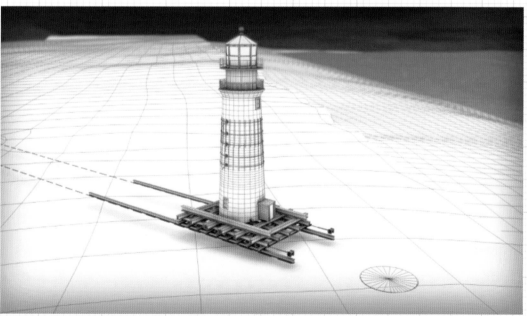

Top Movers suspended the iron staircase inside the structure from the upper floors using steel cables.

Above The lighthouse was moved away from the bluff edge on steel rails. The rails were leapfrogged in front of the structure as its journey inland progressed.

RESCUING NEW ENGLAND'S BRIGHTEST LIGHT

It may not be a tall lighthouse, but moving Block Island Southeast Lighthouse on Rhode Island off the edge of a crumbling cliff ledge proved a formidable challenge. Completed in 1874, the stunning Victorian Gothic redbrick structure originally sat over 200 feet (60 m) away from the edge of the Mohegan Bluff. But over a century of erosion brought the building to the edge of destruction. By 1993, only 55 feet (17 m) of land stood between it and the cliff edge. The brightest light on the Atlantic seaboard was in danger of being lost forever.

Islanders soon raised the funds to help rescue their beloved lighthouse. A team of engineers from International Chimney in Buffalo, New York, and the U.S. Army Corps of Engineers joined forces with Jerry Matyiko's crew at Expert House Movers to haul the 2,000-ton structure over 300 feet (90 m) inland to safety. The crews had to ship all their equipment and machinery out to the island by barge. This was a huge task, as the gear list included over 400 tons of steel beams, 38 lifting jacks weighing 60 tons, huge stone-cutting machines and over 80,000 bricks.

The unique challenge with moving this landmark lighthouse was to prevent its 52-foot (16 m) tower from detaching from the adjoining three-story keeper's house. The team had to tension the steel lifting beams underneath the lighthouse precisely to compress the tower ever so slightly in toward the house to keep the two structures together during the operation.

A team of 25 men toiled through the summer of 1993 to undertake the move. Weather and sea conditions around the exposed cliff ledge changed minute by minute. Fog rolled in so thickly sometimes that the movers lost sight of the lighthouse. "It was so thick you could barely see your hands in front of you," recalls Jerry Matyiko.

On 11 August 1993, four ram jacks, christened Jerry, John, Joe and Jimmy after the four Matyiko brothers leading the move, gradually pushed the lighthouse along steel rails away from the brink of disaster. The building traced a three-step shuffle inland as it moved onto safer footings, where it stands proudly to this day.

To insert the steel frame through the building, the team had to cut a series of deep holes in the lighthouse's base. Rick and Jerry's band of men spent 14 days attacking the lighthouse's 3-foot (1 m) thick walls with diamond-tipped chainsaws, filling the sea air with the squealing sound of metal on brick and chewing through vast quantities of tobacco in the process. Eventually, 10 perfectly rectangular holes were dotted around the lighthouse's base. Jerry's cavalry arrived with truckloads of massive steel beams. "Wrap it around the beam, wrap it around the beam," cried worker Mike as the crew lashed chains around the 30-ton beams. Connecting the chains to an ancient bulldozer, they toiled for days threading the beams through the base of the lighthouse to install the frame. It was a precarious procedure. One misdirected shove and the whole building could topple. As Mike and Jerry slowly waltzed the beams into place, a second team laid out a runway of steel tracks along which the lighthouse would roll.

GOING UP!

"I need to concentrate with all these gauges. If one jack messes up, the pressure will rise on the gauge," said Jerry, scanning the bank of dials on his jacking machine. With powerful hydraulic lifting jacks positioned underneath the lighthouse, crowds gathered at a safe distance to witness liftoff, as Jerry raised the big lighthouse off its old footings. It was crucial that the lighthouse remained perfectly level. The slightest lean could tip the building off balance. His plan was to crack the structure up, off its concrete foundation pad at ground level. This would leave the lighthouse with a flat base that could slide easily along the tracks to the new site.

Jerry threw the lifting levers. The jacks tightened with a screeching hiss. The sand around the base of the lighthouse began to tremble and the resounding

Above Steel cables prevent the iron steps of the staircase from unthreading off their support column as the lighthouse is lifted off its footings and moved.

Below Hydraulic pistons positioned behind the lighthouse extend to push the structure along the rails.

MOVING AMERICA'S TALLEST BEACON

Cape Hatteras Lighthouse, North Carolina, is the tallest brick lighthouse in America. This 198-foot (60 m) beacon was a guiding light for Atlantic seafarers for over a century, warding them away from the treacherous underwater ridges that lie just off the coast. Completed in 1870, the iconic structure, painted in black-and-white candy stripes, was originally built on a sandy hill 1,600 feet (500 m) from the water's edge. But over the years, the cape's coastline has been ravaged by erosion. By 1987, the lighthouse stood on the brink of disaster, just 120 feet (37 m) from the shore.

Cape Hatteras Lighthouse's builder, Dexter Stetson, originally designed the tower to sit firm on 600 foundation piles buried 30 feet (9 m) into the sand. Unable to break through the compacted sand, however, he built a "floating" foundation 6 feet (2 m) deep made from granite rock and

layers of yellow-pine timber. If the salt water reached the layer of wood, there was a danger that the timber would rot and lose its strength. Urgent action was needed to protect the building from the ocean, or seawater could undermine the tower's footings and cause it to collapse. A similar crisis had led to the demise of the Cape Henlopen Lighthouse, Delaware Bay, which crumbled into the ocean in 1926. In 1997, the National Academy of Sciences published a report advocating that the best way to preserve Cape Hatteras Lighthouse was to move it 2,900 feet (900 m) inland, away from the water's edge, onto a sturdier concrete, steel and brick foundation. The move would keep the building safe for at least another century.

The project cost $12 million, funded by a government grant, and united teams of geologists, engineers,

conservationists, structural movers, architects and archaeologists from across the country. Moving contractor International Chimney of Buffalo, New York, worked with Jerry Matyiko's team at Expert House Movers to undertake the relocation. The teams' greatest challenge was keeping the 4,400-ton lighthouse level as they moved it inshore – not so easy when the ground was made from soft, rippling sand. The slightest lean could damage the building's walls or cause it to topple.

The team cleared a straight, flat corridor through the sand and dense brush, compacting it with crushed stone and gravel. They abutted lengths of steel beams side by side, to create a reinforced runway to the new footings, and installed a network of sensors inside and outside the building to monitor wind and weather, stress and tilt. Computers – rarely used in building relocation projects – made a special appearance on this job. Engineers positioned two prisms inside the core of the lighthouse –

one at the base and one at the top. Computers monitored signals sent between the two prisms to determine if, and how much, the building was leaning as it moved. Just in case the high-tech system failed, they hung a trusty plumb bob from the top as backup.

The process of freeing the lighthouse from its old footings turned into a mining operation. Workers had to use wire saws equipped with diamond cutting cables to remove over 800 tons of granite from underneath the structure, inserting a sturdy support frame made from interlocking steel beams in its place. The frame would act as a temporary foundation while hydraulic jacks pushed the lighthouse along steel rails to the new site.

The move began with a fanfare on 17 June 1999. Twenty thousand visitors a day descended onto the beach to cheer Cape Hatteras on its way. Progress was careful and slow, moving an average of 125 feet (38 m) a day. Even when the tower advanced a record 355 feet (108 m) on 1 July, the spectators that turned up could barely notice it moving.

Two weeks into the move, the computer systems indicated that the lighthouse was tilting. When the crew tried to compensate for this, the computer suddenly changed its mind, suggesting that the tower had lurched the other way. The computer system was suffering from a glitch. So movers shut it down and relied on the plumb bob as they rolled the lighthouse the rest of the way.

After 22 days on the move, Cape Hatteras finally reached its new base on 9 July 1999, three weeks ahead of schedule. The cement gluing the lighthouse onto its new footings had barely set when it was pummeled by Hurricane Dennis. Gusts raging up to 130 miles per hour (200 kph) blasted the coastline for five days, cracking several of the tower's windowpanes. The movers had made it ... just in time.

Above Sets of steel rollers positioned between the steel frame built around the lighthouse and the rails guide the tower along the tracks.

cry of "Going up!" rang out across the cliff. But just as the lighthouse began to lift off, other shouts filled the air. "It's coming up from the bottom, Jerry," Mike yelled. Instead of cracking off its base, the whole foundation was lifting out of the ground with the main structure. Like a large tooth with the root still attached, the lighthouse now hung in the air with 5 feet (1.5 m) of extra masonry dangling from its bottom.

Jerry had to act quickly. He needed to slice the extra masonry off the base to give him space to insert support beams under the lighthouse. The masonry was thick and made of solid brick. "We're going to knock all this out, dig it out a little more," he said, jumping into the bulldozer and ramming the side of the lighthouse with a steel snout. Sankaty Head Lighthouse shuddered, but the bricks didn't budge. It was time to break out the jackhammers. For the next three days, the site was an orchestra of noise as the men chipped, bashed and sawed away at the thick brick walls, a vicious wind whipping the sand into throat-clawing dust devils.

For islanders, the sight of the men working under a precariously balanced lighthouse proved an entertaining distraction. Curious holidaymakers crossed the island to take photos. Inquisitive local children did classroom engineering projects. Local newspapers urged photographers to catch the drama as it unfolded.

CRAWLING LIKE AN INCHWORM

At 7.30 a.m. on 1 October, crowds gathered on Sankaty bluff. The 400-foot (122 m) path to the new site stretched out like a mud runway. With the jacks primed and beams straightened, celebratory champagne exploded off the lighthouse wall, wishing the building a safe journey. With ceremonial flourish, Jerry pushed the main lever down and two enormous pistons slowly extended behind the lighthouse, pushing it forward. Once the pistons had reached their maximum extension, the lighthouse stopped and the team sucked the pistons back in and repositioned

THE MATYIKOS: KEEPING IT IN THE FAMILY

Big John Matyiko became a house mover by accident. The son of Gabriel and Mary Matyiko, who moved to the United States from Hungary in 1906, John worked on the family farm in Virginia and ran a small land-clearing business. Weighing around 280 pounds (127 kg), he quickly became known locally as Big John.

In 1956, he won a contract from Langley Air Force Base near Newport News to demolish over 20 houses. The job would change his life and the destiny of the Matyiko family forever. While he was undertaking the clearing work, a local woman living in a pitiable house asked Big John if she could have one of the doors off one of the houses for her home. He obliged, but when he came around to fit the door, the woman said, "The house doesn't do the door justice."

Being a generous man, Big John suggested that since the houses were being knocked down, he could haul one of them to her site as a replacement home. He called a local house mover for a quote, but was horrified at the $1,200 price. So he decided to move it himself. Big John House Movers was born.

Big John begged, borrowed and built tools and equipment to get his house-moving business up and running. He collected old railroad irons to use as steel, bought hundreds of hand-cranked railroad jacks to do the lifting, hand built dolly wheels for the moving, and scrounged railroad ties to use as cribbing.

He even cut an old trailer in half and extended its width with steel "wings" to balance houses on top of. Big John's "land barge" became a regular sight, whisking homes around the state. Whenever he heard of an old house that was about to be demolished, he offered to remove it from the land, take it to his yard, then sell it on later, delivering it to the buyer's doorstep.

Business really took off with the construction of the U.S. highway system in the early 1960s. Hundreds of houses needed moving to create space for the network of new roads. Big John traveled the state taking advantage of the moving jobs on offer, growing the business.

Big John and his wife Isabel had four sons — John Jr., Joe, Jimmy and Jerry — and by the late 1960s, all four of them were working alongside their father hauling houses for a living in Virginia. They eventually set up additional house-moving divisions in Maryland and Missouri, changing the name of the family business to Expert House Movers, growing to become the titans of structural moving in America. Between them, the Matyikos move hundreds of structures every year across the country.

No job is too big or too small for the Matyikos. "If man made it, we can move it," says Jerry Matyiko. While the largest, most complex projects such as moving America's iconic lighthouses keep the family challenged and bring them nationwide notoriety, the smaller house moves, which are always emotional experiences for the homeowners, are a large part of the Matyikos' motivation.

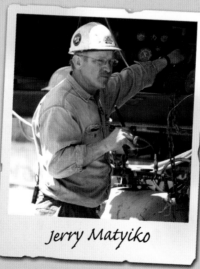

Jerry Matyiko

"Houses in the way of road-widening projects or commercial developments are our stock-in-trade jobs," says Jerry. "But we also get many calls from people whose homes are in genuine jeopardy. A house is a person's greatest investment in life. So if it's about to fall off the edge of a collapsing cliff or sitting in a flood plain full of water, then we'll be there to haul it out of harm's way."

Above The lighthouse moves along the rails away from the cliff, out of harm's way. Around the outside of the building, a support collar made from wooden struts and steel cables holds the brickwork steady.

them before pushing forward again. "It's almost like watching an inchworm crawl," explained Jerry's son Gabe Matyiko, who had been working alongside his father on the move. "As these rams stroke out about 66 inches [168 cm] we stop, we release the jack-in-a-box, suck it back up, lock the jack-in-a-box and go again."

Inch by inch the lighthouse crept away from the cliff. Its slow movement was barely perceptible. But the crowds cheered regardless and Jerry's moustache curled into a broad smile. Over the course of the next three days the lighthouse crawled forward at a snail's pace, rarely moving more than 3 feet (1 m) an hour.

DOWNHILL RUN

While Jerry focused on keeping Sankaty moving, Joe inspected the ground farther down the runway that had softened with a recent downpour. "This terrain is a little bit snaky on us," said Joe. "Parts of it are real good. Other parts are very sandy and loose." The soft sand made life increasingly difficult for the team. "Something's fallen out there," cried Jerry as the lighthouse suddenly lurched to one side. One of the rollers had broken, likely caused by a set of rails sinking into the sandy soil. Jerry's team rallied quickly to right the lighthouse and steer it back on course.

THE LITTLE LIGHTHOUSE THAT VANISHED

Late one night, the eerie sounds of saw blades slicing wood interrupted the sleep of some of the residents in the town of Greville Bay, Nova Scotia. No one was quite sure where the noise was coming from. But the following morning, when residents pulled open their curtains to take in the views of the town's harbor, they were in for a shock. Their beloved lighthouse had vanished!

Port Greville Lighthouse was built in 1908, when the town was the epicenter of Canadian shipbuilding. Hundreds of schooners and fishing boats passed through its waters each week. The wharfs and bustling sea traffic soon made navigating the waters treacherous. So the town's white wooden beacon, shaped like a pepper shaker, was commissioned to keep ships out of harm's way.

By the 1980s, the port's fortunes had changed. Its shipbuilding industry was in decline and the lighthouse was no longer needed. Most residents knew nothing of the Coast Guard's plans to cut the structure in half, load it onto trucks and move it to the grounds of the Coast Guard College 250 miles (400 km) away in Sydney for display. Which explains why they were so stunned that fateful morning to find their beloved lighthouse had disappeared without a trace.

Port Greville's residents launched a search-and-rescue mission to track down their much-missed landmark. After years of campaigning, the building was eventually transported back to Greville Bay. Well-wishers lined the streets to cheer the return of their guiding light. The building was quickly reassembled, repainted and returned to its former glory, positioned on a prominent new perch where residents could keep a watchful eye on it.

THE UNLUCKY LIGHTHOUSE

Not all lighthouses are as lucky as Sankaty Head, Cape Hatteras or Block Island, and escape out of harm's way unscathed. The sad story of Egg Rock Lighthouse underlines how risky it is moving these buildings off the edges of rocky cliffs or receding sandbanks.

Egg Rock is a small 3-acre (1 ha) island off the coast of New England, United States. Buffeted by Atlantic storms, the lighthouse was originally built here in 1856, to ward sailors away from the coastline. The light itself sat on the roof of a two-story, wood-framed house where the keeper lived.

The beacon glowed for over 60 years, before being decommissioned in 1922. The six-room keeper's house was a desirable dwelling. So the government auctioned it off to anyone who was able to haul it off the rock to the mainland. The Egg Rock Lighthouse eventually found a buyer who hired a moving company to slide the house down the cliff face onto a barge to sail it across the water. But just as they were lowering the lighthouse onto the barge, one of the ropes holding the building snapped. Egg Rock Lighthouse lurched forward and fell into the ocean. Waves slowly broke the lighthouse into pieces that washed up on the beaches nearby.

Halfway along the runway, the route changed gradient. "The grade is starting to drop off rapidly and we are going to come down a little over 10 foot [3 m] in height ... so things get a little more complicated," said Joe, surveying the slope. If the lighthouse's rails followed the line of the hill then the building would tilt forward on its descent, putting huge stresses on the outer walls. To keep the lighthouse level, the team built a series of "steps" down the slope. At each step, they transferred the weight of the structure off the jacks onto temporary crib towers made from blocks of wood. They then gradually released the jacks and removed the support blocks to lower the lighthouse. It was a slow and painful process, taking three hours to descend each step before pushing on in leaps of 50 feet

(15 m). "You have to keep your guard up, right till it's down to where it's gonna be for 100 years," said Jerry as he guided the tower down its final step.

On 11 October, Sankaty Head Lighthouse gradually crept over its new footings. "Give me an inch ... a half inch ... just touch it! Whoa ... that's it, folks," cried Jerry. With a weary raise of his hand he stopped the movement and the crowd let out a cheer. Sankaty Head Lighthouse had safely reached its destination.

LIGHTING THE LIGHT

In the weeks after the move, an extensive restoration program breathed new life into the old lighthouse. The protective collar was removed, the tower's masonry restored and its signature red-and-white stripes repainted. A large plaque was laid at the original site to honor those who had donated to the rescue. The light was turned back on and the village of Sconset was reunited with the familiar arm of light that punctuated their evenings.

While the move was a huge success, everybody knows that it is only a temporary reprieve for Sankaty. The houses on Nantucket will continue to fall into the sea, and the islanders will continue to battle the forces of nature in vain. Sankaty Head Lighthouse will be at risk again in around 60 years' time, when another round of fund-raising will be needed to rescue it from oblivion.

All of this is good news for house movers, whose business often comes from people determined to plant their homes in high-risk places. But it forces us to confront difficult questions about where we should and shouldn't build in a changing world. Coastal erosion, fluctuations in sea levels and storm surges look set to increase in the coming century. Not every community has Nantucket's wealth to help rescue it. While the ocean will always provide allure for holidaymakers and homeowners, it is a fierce and unrelenting force with little respect for the fragile boundaries that separate land from water.

Expert House Movers and International Chimney joined forces in a bid to save Nantucket's iconic lighthouse from being swept away by the ocean.

Clockwise from top left
Sankaty Head Lighthouse before relocation, perched on the edge of the eroding bluff; the lighthouse is rolled inland on rails; push jacks thrust the lighthouse forward along the rails; a steel rig erected around the base of the lighthouse supports the building during the move; the lighthouse prior to its move.

Supersize Submarines

THE *ONONDAGA* SAGA
MOVING CANADA'S SUBMARINE

Saving his old submarine had always been Moe Allard's dream. But the mission to transport the fragile hulk of HMCS *Onondaga* to his hometown turned into a challenging eight-month crusade against the ravages of the elements and the ocean.

MOVE STATISTICS	
The *Onondaga* Submarine	
Built	1965
Material	Steel
Weight	1,400 tons
Length	300 ft. (90 m)
Distance moved	600 miles (1,000 km)

Rusting in Halifax harbor, Nova Scotia, sat four historic vessels. Once the pride of the Canadian Royal Navy fleet, the Oberon Class submarines were designed to undertake top-secret surveillance missions during the Cold War. Built in the United Kingdom, these battery-powered "O-Boats" were much quieter than nuclear-powered submarines, making them the silent predators of the ocean. But in 2000, after 33 years of service, the O-Boats were decommissioned and destined for the scrap heap. "What you see here is just basically the hulk of the submarine," explained their minder, Chief Petty Office Robert Arbour, patrolling the submarines from the dockside. "All the batteries, the fuel, all the weapons taken off ... For me it'd be a loss if we didn't preserve one of these submarines at least."

The days for Chief Arbour's fleet looked numbered until former submariner Maurice "Moe" Allard threw him a lifeline. Moe had served on the O-Boats for nearly two decades, and was passionate to preserve one of these iconic vessels. "The Oberon Class of submarines were the best submarines in the world at the time," he explained. "They were the quietest submarines because their main function was – to listen – to come unseen and unheard. We'd sail in the middle of the night and come back in the middle of the night – people didn't know where we were or what we did."

SAVING THE *ONONDAGA*

To commemorate the O-Boat missions, Moe and a team of volunteers in the Eastern Quebec fishing town of Rimouski raised over $800,000 to realize an ambitious plan to build the country's first submarine museum. Sitting on their harbor, next to the town's iconic lighthouse and maritime museum, would be the centerpiece exhibit – an O-Boat. "This here is going to be the only submarine museum in Canada," said

114

Above Four Oberon Class submarines moored in Halifax harbor

Below Marine engineer Donald Trembley

Moe, surveying the proposed site on the dry, stone shoreline. "It's going to bring people from all over, to see how we lived, how we worked and what life was like on board a submarine."

The O-Boat chosen for a history debut in Rimouski was HMCS *Onondaga*. The vessel was huge, stretching nearly 300 feet (90 m) long, protected by a double hull built from steel and weighing over 1,400 tons. After years of neglect, she was also extremely fragile. Transporting the submarine over 600 miles (1,000 km) from Halifax to Rimouski would be a challenging operation.

Moe entrusted the move to veteran sea dog and marine engineer Donald Trembley. Donald was one of Canada's most formidable maritime adventurers. In 1964 he had discovered Canada's largest wreck – the *Empress of Ireland* – which sank not far from Rimouski. The treasures and artifacts that Donald salvaged from the shipwreck were used to found Rimouski's maritime museum. Forty years on, Donald was now hoping that the *Onondaga* would be his final gift to the museum that he had helped to establish. "It's a big technical challenge. It's a project of love ... this is once in a lifetime," said Donald as he prepared to launch the ocean odyssey.

AMBITIOUS PLAN

Work on the operation began during the summer of 2008. The move would take place in two stages. First, Donald's team would have to tow the *Onondaga* from Halifax to Rimouski harbor. Then, they would have to devise a way to move the vessel up the seashore onto the rocky beachfront. With no dry docks or heavy-lift cranes in Rimouski, Donald needed to harness the power of the summer's highest tide to propel the submarine as far up the beachfront as possible, pushing her onto a makeshift railroad track designed to help guide the vessel up the shoreline. Winch trucks attached to the front of the submarine would then pull her up the tracks to her final resting place, leaving her high and dry.

This was a hugely ambitious plan without precedent. For Donald's scheme to work, he would need a tide at least 15 feet (4.6 m) high. There was only one time during the whole of the summer when the tide rose this high – for two short hours in the middle of the night on 2 August. This was a deadline Donald could not afford to miss.

TREACHEROUS TOW

Before the *Onondaga* could embark on her journey up Canada's East Coast, Chief Arbour's team had to make her safe to travel in open water. The rough seas could toss and turn the powerless submarine in the ocean. If she were to capsize, there was a real risk that she could drag the tugboat down with her, so it was vital that the submarine sat as level as possible in the water. But floating in the harbor, the O-Boat was listing badly. When the *Onondaga* had been decommissioned, the vessel's arsenal had been removed along with 18 tons of batteries. This had left the bow of the submarine riding much higher in the water than the stern. The team spent two days pumping 40,000 gallons (180,000 l) of water into the submarine's ballast tanks to level up the ship for the tow. Inside, they sealed the doors to every compartment to ensure that if the *Onondaga's* rusty hull sprang a leak, the water would be contained and would not flood the entire ship.

Below The O-Boat lashed to the tug as she leaves Halifax to embark on her journey to Rimouski

While Chief Arbour's crew readied the *Onondaga* for her treacherous tow, in Rimouski Donald's team rallied to build a "marine railroad" to hoist the *Onondaga* up the rocky shore onto dry land. The crew called in bulldozers to create a smooth ramp up the shoreline, removing any protruding rocks that could puncture the submarine's rusting hull. Progress was slow. The team had to down tools each time the tide washed in, flooding the worksite. The move suddenly became a real race against time. "We have to be very fast in our operation and we don't have much time," said Donald as the diggers scurried around on the beach. "The tide will not wait for us. If we miss it – it's a catastrophe."

THWARTED BY FOG

On 9 July 2008, the *Onondaga* was ready to leave Halifax and embark on her final voyage. Tug captain Joseph Legge was handpicked for the operation. With 40 years' experience, he had survived more than his fair share of life-or-death adventures working on the high seas. In 2006 Joe had been awarded a medal of bravery for saving 13 fishermen from a burning ship during a heavy Atlantic storm 45 miles (72 km) off the coast of Newfoundland. Despite his track record, the task of towing the *Onondaga* to Rimouski made him extremely nervous. "I didn't sleep last night. This is like launching the space shuttle. Anything could happen. We could roll over, spring a leak, break the tow wire … you never know what could happen."

Below It was crucial that the submarine sat level in the water.

HOW TO SAIL A SUBMARINE ON LAND

Hauling the *Onondaga* from the water onto the dry harbor front was the hardest part of Donald's mission to deliver the vessel to Rimouski. With no docks or cranes to lift the submarine out of the water, Donald built a marine railroad 330 feet (100 m) long up the shoreline to the harbor front. Five cradles, shaped to support the underside of the submarine's hull, sat on top of the tracks.

To propel the submarine into the cradles, Donald's team needed a high tide of 15 feet (4.6 m). The only time during the summer of 2008 when the tide rose this high was during the night of 2 August.

To steer the vessel onto the cradles with precision, and to prevent winds or waves pushing her off course, Donald's team tethered the submarine to two trucks on the pier and two buoys in the sea using ropes. By tightening and loosening the ropes, and moving the trucks along the pier to shift with the waves and tide,

the team could guide the *Onondaga* steadily up the shoreline as the tide pushed her onto the cradles.

It was vital that the strongest parts of the submarine's hull – her bulkheads – lined up precisely over the cradles to ensure that the vessel was balanced along the length of the rails and to avoid her toppling over as the tide retreated. At high tide, once the submarine was engaged in the cradles, three winch trucks attached to the front of the vessel with over 2,300 feet (700 m) of rope would pull her up the rails.

The cradles would act like trolleys and allow the submarine to roll smoothly up the rails, keeping her on track on her journey up to the harbor front.

Once the submarine reached her final location, she would be surrounded by a protective bank of rocks and secured in place with firm supports. A gangway would allow visitors to venture inside.

THE SUBMARINE BUILT BLOCK BY BLOCK

Building a vessel as large and complex as a submarine from the ground up is a long and technically demanding process. To streamline the construction of the British Royal Navy's fleet of new Astute Class nuclear submarines, teams of engineers at BAE Systems build the vessels in separate modules and then move each completed piece to a giant assembly hangar where they are fused together. This form of "modular construction" involves dividing up the submarine into distinct sections such as the engine room, nose and command deck. Teams of engineers can work simultaneously to assemble each module in a different hangar, reducing the vessel's overall construction time.

Moving each module to the central assembly hangar is a unique challenge. The team has to load some of the units, built in Barrow-in-Furness, onto multi-wheel transporters and guide them through the small town to reach Devonshire Dock Hall, where they are joined together. Constructed at a cost of $400 million, the assembly hall soars 12 stories tall and houses six heavy-duty cranes. The building is so vast that two submarines can be assembled alongside each other.

Each completed submarine weighs around 7,400 tons and is the length of a football field. To ensure that they reach the water intact, the vessels are assembled on top of cradles. The cradles sit on top of rail cars that run along tracks to guide the submarine out of the hangar, along the dock and into the yard's giant ship lift at a speed of around 3 feet (1 m) every five minutes. The ship lift then lowers the submarine into the water, where she can be tested before full deployment.

A heavy blanket of fog hung over the harbor that morning. The outlook for the journey didn't look good. "We got up this morning like 3.30, to get the new forecast and the forecast wasn't good," explained Joe. "The problem is, we don't know what we're dealing with. We've never towed anything like this before. Leaving on a bad forecast is not a good idea." Joe could not predict how long the journey to Rimouski would take. If they arrived later than 2 August, they could miss the high tide. But Joe didn't want to risk leaving with such a dire forecast, so decided to delay departure.

Above The O-Boat en route to Rimouski

THE LAST VOYAGE

The skies had finally cleared by the morning of 11 July. With rays of dawn sunlight beaming down onto the *Onondaga*, Joe cast off and set a course for Rimouski. "It's a relief to finally see it under way after the delays we've had for the last two or three weeks trying to get her going," remarked Chief Arbour as his boat set sail for one last adventure. "Hopefully they'll get her up there without any problems."

But no sooner had Joe pulled the *Onondaga* away from the docks into the middle of Halifax's harbor than disaster struck. Under intense pressure, one of the tow bridles attaching the submarine to the tugboat snapped with a loud clatter. Cast adrift on a turning tide, the submarine flailed around, threatening to block marine traffic in Halifax's busy harbor. With the operation dangling by a thread, Joe managed to wrestle the O-Boat back to the dockside and tie her down again with the help of a backup Navy tugboat. "I'm glad it happened there in the harbor and not out at sea," said Joe. "It doesn't want to leave its buddies behind tied up over there. Bad start to the day," said Chief Arbour.

The team spent the rest of the day replacing the tow bridle with a much thicker chain. After the six-hour pit stop, they launched again. This time, the *Onondaga* sailed smoothly out of the harbor into the open ocean. "Everything looks good," chirped Joe at the wheel as the trip got under way.

HURRICANE AHEAD

The next day, 200 nautical miles (370 km) from port, Joe received distressing news over his radio: a storm was chasing their tail up the East Coast. Hurricane Bertha could easily upset the precariously balanced load if it closed in on them. "You got a ... big huge submarine like we have here, this fire-breathing dragon breathing down your neck, waves can reach over 25–30 feet [7.5–9 m] in a matter of a few hours. She'll capsize on you," said Joe nervously. To avoid being consumed by the storm, Joe plotted a diversion inland, through a narrow strait on the southern end of Cape Breton Island. This route would protect them from the storm, but involved navigating the huge O-Boat through a series of narrow locks.

No submarine had ever traveled through this particular canal before. Concerned that the O-Boat's fragile hull might be damaged if she hit the locks' concrete walls, the team called in a second tugboat to help steer her through. Joe's boat tugged the front of the O-Boat while the second tug pulled the rear of the vessel tight. This inventive maneuver kept the submarine straight as she slowly squeezed her way through the locks.

Below Tugboats at the front and rear of the O-Boat pull tight on the vessel to keep her straight as she enters the locks.

RAISING THE *KURSK*

When the Russian Navy Oscar II Class submarine *Kursk* suffered explosions and sank in the Barents Sea during a training exercise in August 2000, the task of salvaging the vessel to recover the bodies of the 118 crew killed on board was a highly complex and sensitive operation. The damaged vessel contained two potentially unstable nuclear reactors, along with an arsenal of unexploded missiles and torpedoes.

The accident had crippled the front bow section of the submarine where the torpedo room was situated, rendering it extremely fragile. If engineers attempted to crane the submarine out of the water with the bow attached, there was a risk that it could separate from the vessel, putting a sudden strain on lifting equipment. So a consortium of experts from Netherlands-based engineering firms Mammoet and SMIT International devised a scheme to separate the damaged bow from the central hull using an underwater "ship saw" before they raised it to the surface.

Before the salvage work could begin, engineers removed the hazardous munitions. Radiation levels around the hull were measured and recorded to ensure they were safe for divers. A long metal cutting cable was strung over the vessel's bow, just behind the bulkhead of the torpedo room. Two suction anchors, positioned on the seafloor either side of the submarine, guided the cable. Winches pulled the cable back and forth, enabling it to gradually slice down through the vessel's double hull, cutting away the damaged bow.

With the bow removed, divers drilled a series of holes in the submarine's hull and attached 26 computer-controlled lifting lines to her body. The lines were linked to strand jacks on SMIT's transportation barge, *Giant 4*, on the surface.

Over the course of 15 hours, the jacks slowly raised the rear half of the submarine up off the seabed onto supports fixed underneath the barge. The jacks were fitted with special compensators that acted like shock absorbers to counteract movements of the barge caused by waves, to keep the submarine level as she was lifted.

The *Kursk* was transported attached to the underside of the *Giant 4* to a dry dock in Murmansk. The successful salvage mission enabled the bodies of the crew to be recovered for burial and provided a chance for investigators to probe what caused the disaster.

Top Suction anchors on the seabed guide a metal cutting cable to slice the bow of the *Kursk* away from the hull.

Above Lifting lines were used to raise the *Kursk* off the seabed.

Above The vessel is raised out of the water in Murmansk.

Above The O-Boat arrives in Rimouski.

While the *Onondaga* made her way up the coast, Donald's team completed work on the 330-foot (100 m) marine railroad in Rimouski ahead of the crucial 2 August deadline. As the *Onondaga* pulled into Rimouski's port, thousands of locals flooded onto the wharf to get their first look at their new submarine. This once-secret sea predator was now the talk of the town. It was a glorious day for Donald and Moe as their dream of landing an O-Boat on Rimouski's harbor front came tantalizingly closer to becoming a reality. "It's a milestone for us here," said Moe, overjoyed at the vessel's safe arrival. "I guess the whole town dropped down to see it. The majority of them have never seen a submarine!" he said as onlookers crowded to take a closer look. Both Moe and Donald knew, however, that the most challenging part of the mission was still ahead.

A STORM STRIKES

On the morning of 2 August, Donald's team made their final checks on the railroad track, ensuring that the cradles that would support the weight of the submarine as she rolled up the tracks were aligned with the rails. They were pinning their hopes on the 15-foot (4.6 m) high tide, which would peak just after midnight, pushing the submarine onto the cradles, so trucks could winch her up the rails to the harbor. But a storm was stirring out at sea.

"The wind looks like it is changing," said Donald as dark clouds brewed overhead and gusts rolled in, churning massive white-capped waves on the St. Lawrence. "This isn't a good solution," said Donald as the skies blackened and rain started to pour down. He faced a difficult decision. In such torrential weather, the submarine could easily careen out of control in the sea, puncturing her hull on the tracks or the shoreline. But if he canceled the operation tonight, he would miss the only 15-foot (4.6 m) high tide of the summer. The rain lashed down on the crew, washing away any enthusiasm they had for the mission. At midnight, Donald called them to an emergency meeting inside one of the on-site cabins. "Midnight's too late. Forget it, it's finished," Donald said crisply, calling off the move. His crew couldn't believe his call. "Is it really canceled?" they asked, stunned. "Pack up and go home. It's all canceled," reaffirmed Donald.

A NEW PLAN

The crew couldn't afford to leave the *Onondaga* floating in Rimouski's harbor into the winter. If the water froze over, the ice could crush her fragile hull. They had to revise their plans and pull the submarine into shore on a lower tide. The crew called in underwater welders to fuse additional lengths of track together, extending the railroad deep into the water. Donald hoped that this would enable them to work with a tide 14 feet (4.3 m) high.

A month later, the crew gathered on the shoreline for a second attempt. As the sun set, the sky remained clear and the water eerily calm. The omens looked good. At midnight Donald lined up the trucks on the harbor pier, roping them to the submarine to help line her up with the rails. He fastened additional winch trucks to the front of the submarine to help pull her up the rails.

Donald assumed command from the top of the submarine's conning tower. With a CB radio clutched in his hand and a clear view of the scene, he called orders to his team. Donald had no way of seeing the underside of the submarine's hull to check its alignment with the rails, so he had set up a series of metal guideposts on top of the vessel. These had to align perfectly with additional posts positioned on

Below Ropes anchor the side of the O-Boat to trucks on the pier. Steel cables on the bow are tensioned to help guide her up the rails inland, out of the water.

THE SHIP THAT MOVED A SUBMARINE

Supersize submarines are occasionally hauled onto oceangoing ships for speedy transfers. In 2003, following service in the Persian Gulf, rather than sail the submarine HDMS *Sælen* back to Denmark, the Royal Danish Navy opted to piggyback her on a cargo ship to return her home more quickly. The vessel chosen for the job was the German heavy-lift cargo ship *Grietje*. This 500-foot (150 m) ship, operated by SAL Shipping, has three cranes on her deck capable of lifting hundreds of tons.

Marine engineers used tugboats to steer the *Sælen* as close to the *Grietje* as possible to enable the cranes to lift the load with maximum power. Divers in the water helped guide the submarine into custom-built cradles attached to the cranes' lifting cables. The cradles sat underneath the *Sælen* to support her 370-ton hull and keep her level. The team had to coordinate the movement of the cranes to ensure that they lifted and swung the submarine down onto the deck of the *Grietje* in unison.

With straps holding her tight on the *Grietje*, the *Sælen* made a speedy return to Denmark. The *Sælen* was decommissioned in 2004 and today sits on display at the Royal Danish Naval Museum in Copenhagen.

the shore to help guide the submarine in. From Donald's vantage point, these posts worked like the sights on top of a gun. When the posts lined up, he would know that the submarine was aligned perfectly with the rails. As the tide levels peaked, everything seemed to be going perfectly to plan. But just as Donald was lining up the posts, a freak cloud of fog rolled in from off shore. The fog was so thick that Donald could no longer see the front of the submarine or the posts on shore. "I don't see my alignment. There's too much fog," he radioed to the crew in the water. "I'm screwed with the fog," he said, exasperated.

Donald had to make his best guess on the submarine's alignment and signal to the winch trucks on shore to begin hauling the vessel up the tracks. If they waited much longer, the tide could pull her back out to sea. Three winch trucks rumbled into action. Donald had rigged the cables through a block and tackle to increase their pulling power. With the might equivalent to three steam locomotives, they began tugging the colossal submarine out of the sea. Moe smiled as he saw the belly of the beast clear the water, her hull crusted with barnacles. "It's like an orchestra," he said, marveling at the might of machinery working together to wrestle the *Onondaga* onto dry land.

By the time the tide began to turn, the team had brought the submarine halfway home up the tracks. The retreating tide made it impossible for them to pull the O-Boat any farther up the shore. At 3 a.m., Donald and his team retired for the night, pleased with their progress. They would return tomorrow for the final pull up the rails. But while everyone was asleep, disaster struck again.

STRANDED!

The following morning Donald and Moe arrived at the harbor to a heartbreaking sight. The *Onondaga* had overturned and fallen off the tracks. She was lying on her side on the shoreline like a beached whale. Tears welled up in Moe's eyes as he surveyed the scene. "It's sad. It's

Above As the tide retreated, the O-Boat keeled over and fell off the tracks.

almost like losing a baby," he said. Donald immediately examined the wreck and located the cause of the accident. The submarine had been misaligned over the cradles. When the water had receded and her full 1,400-ton weight was transferred onto the cradles, they had become overburdened on one side and given way, causing the submarine to roll off the tracks. To make matters worse, a large boulder had punctured a hole through the submarine's outer hull. Fortunately, the inner hull had remained watertight.

There was a chance that Donald's team might be able to refloat the *Onondaga* on another tide before the winter. So over the next two weeks, they raced to replace the mangled cradles and right the submarine using hydraulic jacks. This time, high tide fell during daylight hours. They used two tugboats to help guide the submarine into line with the tracks, but just when it looked like they might succeed, an overanxious tugboat captain pulled the *Onondaga* too hard to one side, causing her to tumble over into the sea. Moe was devastated. "I was positive we were going to have it today," he said. "She's a stubborn old lady."

RESCUE MISSION

With his reputation at stake, Donald remained resolute. He and his crew worked tirelessly into the winter to lift up the battered submarine again for one final push. On 28 November, they lined up the trucks, boats and winches as snow began to settle on the submarine's hull. In a matter of days, sea ice would engulf the ship. This would be Donald's final chance to haul her out of harm's way onto dry land. "Just in time for Christmas," said Donald as he assumed control of the flailing vessel. Luckily for Donald, Moe and the team at the museum, the final pull was a slow but sure triumph. As snow squalls and driving winds howled across the harbor, the *Onondaga* lumbered up the beach toward her final resting place, cheered on by

Below The O-Boat finally makes it out of the water onto dry land as winter storms move in.

crowds bold enough to brave the freezing conditions to watch the spectacle. Donald and the crew celebrated with champagne as they beached the submarine, high and dry.

Over the following months, they repaired and repainted the old vessel inside and out, reinstalling her masts and periscopes in time for the grand opening of the submarine museum on 29 May 2009. Thousands turned out for the opening ceremony, including 50 former submariners, many of whom had served on the old O-Boat. In the first year, over 100,000 visitors came to explore HMCS *Onondaga*. So inspirational and successful was her epic trek that plans are under way to rescue a second O-Boat and transform her into Canada's second submarine museum.

Above The team celebrates the mission's completion.

Hauling the fragile hulk of the submarine *Onondaga* from Halifax to Rimouski was a hugely ambitious scheme.

Clockwise from top left
The *Onondaga*; the team haul the O-Boat up Rimouski's shore on rails; the team had to work through the night to catch high tides to help propel the submarine up the shoreline; the *Onondaga* in her final resting place in Rimouski, restored and opened to visitors; the team inspect the overturned vessel; righting it by hand wasn't an option!

Aircraft Airlifts

SPECTACULAR SPITFIRE THE FLYING JIGSAW

Aviation enthusiast Tom Blair had dreamed of owning and flying a classic Spitfire aircraft ever since he was a young child. When he finally located a vintage Spitfire for sale in the United Kingdom, moving it across the Atlantic to his hangar in Maryland in the United States proved a meticulous operation for a British team.

MOVE STATISTICS	
Supermarine Spitfire Mk IX	
Built	1943
Material	Aluminum
Weight	3.5 tons
Height	12.6 ft. (3.8 m)
Length	31 ft. (9.5 m)
Wingspan	36.7 ft. (11.2 m)
Distance moved	4,000 miles (6,400 km)

The Supermarine Spitfire is perhaps one of the world's most famous fighter aircraft. An icon of British engineering, the plane helped to turn the tide of the Second World War. The Spitfire cast a spell over British-born aviation enthusiast Tom Blair when he was a child. "As many young boys, I had an interest in planes and built model airplanes. Looking through a magazine one day I saw this beautiful plane with an elliptical wing. It was an absolutely stunning plane," recounts Tom, who lives in Maryland and owns a collection of vintage aircraft.

The Spitfire played a decisive role in many raids, including the Invasion of Normandy, a battle Tom's own father fought in. Tom's father was an American serviceman stationed in England during the Second World War and married while on secondment. A couple of weeks before Tom was born, his father was killed in the Invasion of Normandy. This Anglo connection fueled Tom's fascination with British engineering and history. "As a young teenager I started to understand the significance of the Spitfire in the Battle of Britain and the significance of the Spitfire really to the British people," he recalls.

Today, Tom has lost none of his childhood passion for aircraft. A successful business entrepreneur, he invests much of his wealth in rescuing, restoring and flying old war birds. His collection includes a P51 Mustang, a P40 Curtis, a Czech-built L39 jet, a Beechcraft T-34 and a rare First-World-War Tiger Moth biplane. Maintaining just one of these aircraft in flying condition can cost tens of thousands of dollars a year. Tom has restored all of his planes to working order and employs a crew of specialist mechanics to maintain them. Were it not for his preservation efforts, these aircraft's engines would have seized up long ago and they would be fit for little more than display.

Right Tom's 1943 Spitfire
Mk IX in flight

One aircraft that Tom wanted to add to his collection more than any other
was an original Spitfire. "I don't think there's been an aircraft in the history of the
world that has such a unique combination of attributes," said Tom. "It is absolutely
stunning to look at. It performs beautifully. Flown by valiant British fighter pilots, it
turned the tide in 1940 ... and really, many people might say, helped to save
a nation."

BIRTH OF AN ICON

The Supermarine Spitfire was conceived by British engineers on the eve of the
Second World War as a nimble fighter to intercept enemy bombers. The British
government sent out specifications to aircraft designers, calling for a single-seat
four-gun fighter that was highly maneuverable and capable of reaching speeds of
250 miles per hour (400 kph). In 1938, R.J. Mitchell, the chief designer at Vickers
Supermarine, drew up plans for a prototype called K5054. This revolutionary feat of
engineering would come to be known as the Spitfire.

Below Spitfires flying in
formation at a Battle of
Britain air show at Duxford

The first Spitfire Mk I rolled off the production line and was
delivered on 6 August 1938 to the No. 19 squadron at RAF
Duxford. Today, not far from its old home, a group of aviation
experts and mechanics are dedicated to restoring and maintaining
these classic aircraft. The Aircraft Restoration Company (TARC)
started out as a small team of enthusiasts who restored a derelict
Bristol Blenheim aircraft back to flight. Over the years, they grew
to become one of the world's leading restorers for a wide range
of vintage aircraft, including the Spitfire. Led by chief engineer
and test pilot John Romain, they can transform a wreck into an
airworthy war bird in just a few years. The group's pilots also
regularly mount breathtaking aerial displays, flying up to 16
Spitfires in formation to crowds of thousands.

A FIGHTER REBORN

In 2003, Tom heard that a Spitfire Mk IX was being put on the market at Duxford for around $2.4 million. He immediately called John, booked his flight, and in a matter of weeks was at the hangar needing little persuasion to open his checkbook to purchase the aircraft. "I went over, looked at it, and absolutely was in awe of the plane, and sat in it and could not believe how fortunate I was," recounts Tom.

More than 20,000 Spitfires were built but today only 50 of these aircraft survive to fly. Tom invested $400,000 to have his Spitfire fully restored to working order. Once the work was completed, the team faced the precarious challenge of transporting the vintage plane 4,000 miles (6,400 km) across the Atlantic to Tom's hangar in Maryland. The move would cost Tom nearly $50,000. To deliver the aircraft, the team would have to dismantle its fuselage and squeeze it into a shipping container, sailing it across the ocean in a cargo ship. TARC's aviation expert Martin "Mo" Overall was in charge of the move. Having restored many Spitfire aircraft, Mo had expert knowledge of the pitfalls of dismantling and reassembling these vintage fighters. "They're awkward to hold and they're awkward to handle, and they're very heavy!" he explains.

Below Tom's 1943 Spitfire Mk IX on the tarmac at Duxford

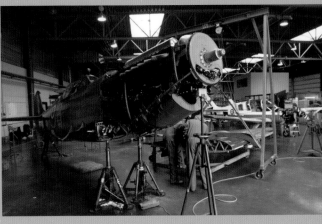

Above, left to right The Spitfire during disassembly at Duxford. After all the major components have been removed, the fuselage is secured into an A-frame with the wings resting on its side.

To fit the aircraft into the container, Mo's master plan involved breaking up the Spitfire into its five major components – the two wings, the tail, the propeller and the fuselage. It would be a delicate procedure. Removing the wings from the fuselage would take away the aircraft's only form of support – its undercarriage – leaving the plane immobile. The wings were also designed to balance the Spitfire. Without them, the weight of the engine and propeller at the front end of the aircraft would render it nose heavy. To prevent the Spitfire from toppling over as they took it apart, the team propped up its underside with support towers. Stripping back the aircraft's panels, they carefully loosened the nuts and bolts that held it together. As they removed its major parts, they checked, cleaned and wrapped every component, bagging and labeling all the fixtures to ensure that each piece could be put back in exactly the same place.

The team had to take extra care lifting the Spitfire's heavy propeller blade off its shaft. "It would be disaster if we dropped the propeller!" said Mo as they called in a forklift to help support its 450-pound (200 kg) weight. "The propeller is driven by a big spline shaft and as it's being drawn off, you have to be very careful to slide it off evenly. If you knock the spline, worst-case scenario, you could shake the engine from the frame," explained Mo. They cushioned the propeller's touchdown on a

soft pad (left) and erected a gantry crane over the wings to support their weight as they lowered them to the ground. It was a delicate balancing act. If the wings were dented or bent, they might not be able to withstand the extreme forces when the plane returned to flight.

Protecting the fragile fuselage in transit was a major concern for John. After removing the tail, the team erected a custom-made metal A-frame around the fuselage to support its weight and restrict its movement in the container. "You run the risk that every container is going to be banged or knocked at some point in the shipping process. There's the road transport, ships, the cranage. I mean the worst-case scenario is that the container actually rolls over and then you really are into a lot of damage," explained John.

STORMY SEAS

With the wings firmly secured on the side of the A-frame, they loaded the aircraft into the container and packed it tightly for the voyage. The container was trucked to Thamesport, 35 miles (56 km) east of London. Thamesport is one of the fastest-growing ports in the United Kingdom, with eight huge cranes capable of lifting tons of cargo every hour. Carving through the water, the 964-foot (294 m) *Hoechst Express* cargo ship pulled into the docks for loading. It would sail the Spitfire along with hundreds of other containers to America.

Operations Manager Neil Monroe ensured that everything ran smoothly on the day. Loading containers onto a ship is not without risks, especially for such delicate cargo. "If the container did drop there would be dramatic damage inside. The bounce alone would probably damage the airframe itself. So it does have to be treated with kid gloves," Neil said.

There were more potential hazards out at sea. Ships crossing the Atlantic are frequently buffeted by stormy swells, and containers stacked high on the decks of cargo ships occasionally tumble overboard. The team had to do everything in their power to ensure that Tom's precious Spitfire wasn't lost at sea. "Due to the value of the aircraft we have requested lower-deck stowage on this particular container," explained Neil. "However it's not guaranteed. All we can do is request it."

Below The A-frame packs tightly inside the shipping container to protect the components of the aircraft from being damaged in transit.

HOW TO DISMANTLE A SPITFIRE

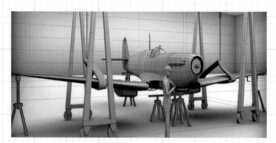

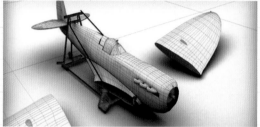

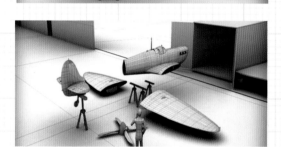

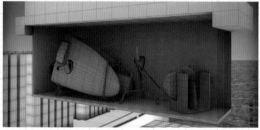

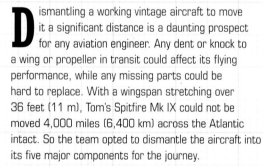

Dismantling a working vintage aircraft to move it a significant distance is a daunting prospect for any aviation engineer. Any dent or knock to a wing or propeller in transit could affect its flying performance, while any missing parts could be hard to replace. With a wingspan stretching over 36 feet (11 m), Tom's Spitfire Mk IX could not be moved 4,000 miles (6,400 km) across the Atlantic intact. So the team opted to dismantle the aircraft into its five major components for the journey.

Before removing any of its parts, the team braced the underside of the aircraft with support towers. They removed the Spitfire's propeller first. Weighing around 450 pounds (200 kg), the propeller is the aircraft's most expensive component. The team loaded the propeller onto a custom-made transit stand to prevent it from knocking into the other pieces in transit. They raised the landing gear and used a gantry crane to support the half-ton weight of the wings while they removed the bolts that secured them to the fuselage. They lowered the fuselage into a steel A-frame and secured the wings onto its side. They protected the tail inside a wooden cradle that had been carefully tailored to fit its shape. The frames and cradle held the components rigid and prevented them from shifting around inside the container during transit.

It took the team three days to put the Spitfire back together in Maryland. When pilot John Romain took the Spitfire up for its first test flight after shipping, the trip proved a little bumpy. On touching back down, he noticed that the Spitfire's right aileron needed readjustment; it wasn't quite flying straight. So John used a block of steel and a mallet to carefully straighten out the trailing edge of the wing to correct its performance. "It's like tuning an instrument," said John, gently knocking the aileron back into shape.

MOVING THE "SPRUCE GOOSE"

Howard Hughes' enormous "Spruce Goose" aircraft has traveled farther on the sea and open road than in the air. Conceived in 1942 to transport military personnel and supplies across the Atlantic, Hughes' H4-Hercules was built from wood, to save precious metal resources for the military campaign. This led to the aircraft becoming nicknamed the "Spruce Goose," even though it was built from birch.

With a fuselage 220 feet (67 m) long and a wingspan 318 feet (97 m) wide, the aircraft is the largest wooden airplane and flying boat ever built. It was not completed until after the Second World War. During the summer of 1946, the airplane's main components were transported to Los Angeles' Terminal Island dry docks, where the aircraft was assembled for test flights. The docks were 28 miles (45 km) away from Hughes' hangar and Culver City stood in the way of the aircraft's huge hull and wings. Hughes'

company called in heavy-hauling specialists Star House Movers to lug the pieces of the aircraft to the docks. The move became a national event and 100,000 spectators lined the city's streets to watch the flying giant make its way to the water.

It took months to assemble all the aircraft's components on the dockside, but in November 1947 the completed plane finally taxied along the water in the bay for testing. It took to the air on the third run, managing to fly 70 feet (21 m) above the water for a 1-mile (1.6 km) stretch before touching back down on the sea.

The flight silenced critics who had poured cold water on the mercurial billionaire's ambitious aircraft, but the plane never flew again. The aircraft sat in a temperature-controlled hangar on Terminal Island until it was dismantled and moved in four giant sections to Evergreen Aviation Museum in Oregon, where it was reassembled for display.

Right The *Hoechst Express* pulls into the Port of Virginia after 10 days at sea.

Right The *Hoechst Express* pulls into the Port of Virginia after 10 days at sea.

It took 12 hours to load all the containers onto the ship for the voyage. The Spitfire won a coveted position deep inside the hold of the *Hoechst Express*, where it would be protected from the ravages of the ocean. The ship set sail into the Medway Channel bound for American soil. It steered a safe course around the dangers en route, including the sunken remains of the U.S. Liberty ship SS *Richard Montgomery*, which became grounded on a sandbank in 1944 while transporting 7,000 tons of munitions to France. The ship eventually sank to the bottom of the Thames estuary, but at low tide her mast is still visible, so the channel's navigational services directed the *Hoechst Express* safely around the wreck as they do for all ships that pass through its vicinity.

NEEDLE IN A HAYSTACK

After a 10-day sea voyage, the *Hoechst Express* was guided by two tugboats into America's oldest port: the Port of Virginia. Port worker Joe Harris patrolled the dockside primed to intercept the precious Spitfire on the busy quay. Despite being armed with the vessel's loading plan, spotting the Spitfire's blue cargo box in the rainbow of containers streaming off the ship's decks was not easy.

"The box we're looking for, we believe it's this row right here, third or fourth from the bottom," said Joe, studying the matrix of boxes on the grid and trying to tally them to the ship in front of him. "That box is probably two or three down, so there'll probably be half a dozen, maybe 10 moves, then they'll get to the box with the Spitfire in it ... so we're getting very close." It was like looking for a needle in a haystack as the boxes were craned off and stacked on the dockside forming multicolored skyscrapers. But after an anxious wait, Joe spotted the Spitfire's container and ensured it was transferred safely onto the truck that would whisk it to Tom's hangar in Maryland.

THE CONCORDE THEY
SAILED UP THE THAMES

In 2004, Londoners were treated to the incredible sight of one of the world's most iconic aircraft floating up the capital's river. The fleet of Concorde aircraft was retired from flight in 2003 and several of the aircraft were donated to museums for preservation. Concorde G-BOAA was bestowed by British Airways to the Museum of Flight at East Fortune near Edinburgh for exhibition. Transporting the 203-foot (62 m) retired aircraft by barge from London Heathrow Airport to the museum in Scotland was no mean feat.

With a wingspan stretching more than 82 feet (25 m) wide and its tail soaring over 26 feet (8 m) high, the aircraft was too big to pass beneath several bridges en route. So in a nine-week operation, engineers dismantled sections of its wings and tail in London prior to its epic journey. The Concorde's fuselage was secured onto a low loader and carefully driven through the streets to the banks of the River Thames at Isleworth, where it was loaded onto the 262-foot (80 m) barge *Terra Marique*. The Concorde set sail on 12 April and passed many landmarks on its stately voyage, including Tower Bridge, the Houses of Parliament and the London Eye.

The barge sailed up the East Coast and was greeted by hundreds of spectators at Torness, East Lothian, where the aircraft rolled back onto dry land and was loaded onto a transporter for the final leg of its journey.

The Concorde crawled at walking pace along the A1 to East Fortune airfield, a far cry from the supersonic speeds it was more accustomed to flying at. Two bagpipers greeted the aircraft's arrival. The wings and tail were reattached and today it sits proudly on display in an exhibition hangar.

THE PLANE THAT SWALLOWS PLANES

When in 2010 a team of aviation enthusiasts needed to rescue a stricken Royal Navy Fairey Gannet attack aircraft from a remote airbase on the East Coast of Canada, they called in one of the world's largest cargo planes to airlift it to a maintenance hangar in Wisconsin where it could be restored to working order. The move was a typical job for the Ukrainian crew of the Antonov AN-124, who fly monstrous cargo all around the globe.

The Antonov AN-124 was originally developed in Russia in the 1980s as a military plane to transport tanks and supplies for the Soviet Air Force. Its four engines generate more thrust than the first Russian space rocket, while its tanks hold enough fuel to fill the gas tanks of 2,500 family cars. Huge doors at the front and rear of the 226-foot (69 m) aircraft enable it to swallow enormous loads

weighing up to 165 tons. To facilitate loading, the aircraft can "kneel down" close to the ground. Two extendable legs, known by its crew as "elephant feet," descend from its nose, supporting the aircraft while the front landing gear folds forward. In this kneeling position, the floor of the cargo bay is just over 3 feet (1 m) off the ground. Winches inside the aircraft help pull cargo up ramps into its hold.

The aircraft has been used to transport commuter trains, boats, elephants, locomotives, aircraft, whales and even an ancient obelisk. In 1988, Soviet engineers built a one-off version of the AN-124 extended specifically to carry the Buran Space Shuttle. With a fuselage stretched by 23 feet (7 m), two extra engines and eight additional wheels, the world's biggest cargo aircraft could piggyback the 60-ton shuttle on top of its huge fuselage.

Above The Fairey Gannet aircraft is slowly hoisted into the fuselage of the Antonov AN-124.

Left The Antonov AN-225 – a specially extended version of the Antonov AN-124 – transports the Russian Space Shuttle on its back.

Above right A racing boat is removed from the AN-124's fuselage.

Below right In 2002, the Antonov team hauls a CH-53 heavy-lift helicopter from Germany to Afghanistan.

TESTING TIMES

John, Mo and the team from Duxford flew into Maryland to help piece together this historic puzzle. When they opened the container, to their relief, the Spitfire had arrived with no visible damage. With help from Tom's own mechanics and pilots, it took the team three days to reassemble the aircraft, like a giant model-plane kit. They took particular care reattaching the aircraft's propeller. "It's crucial to get the propeller set up correct so you obtain the correct rpm," explained Mo, threading the propeller back onto its shaft. "Otherwise the engine could overspeed or come loose."

Before its inaugural flight over American soil, the Spitfire's control cables, electronics and hydraulics needed testing. To trial the Spitfire's landing gear, Mo had to make some ad hoc modifications to his hydraulics testing kit to plumb it into the American machinery. Eventually, it fired up. "The down is obviously more important than the up," quipped Mo as the Spitfire's undercarriage flipped down to the ground successfully. "But it's nice to have them go up as well!" With most of the Spitfire's panels back in place, the team rolled the aircraft out onto the tarmac and filled its engine with fuel, antifreeze and water. They needed to test its engines running at 3,000 rpm. They strapped the plane's tail to the ground and wedged a wooden beam in front of its wheels to stop the Spitfire taxiing during the test. "The aeroplane is producing so much power it wants to go flying and we're trying to hold it on the ground," explained Mo as its Merlin engine spluttered back to life.

There's no mistaking the roar of a Merlin engine. Pilots and ground crew from all over the airport heard the calling and soon gathered at Tom's hangar to watch the Spitfire's test. "This is so amazing, what an impressive piece of equipment," remarked an onlooker as the propeller stirred the air around the aircraft. "Fuel's off, power's off and brakes are off! ... All good!!" said Mo after a successful test run.

The team screwed the final panels into place. Some of the screws had distinctive stripes painted on their heads. Matching them up with the right hole and tightening them precisely so that the patterns of paint all lined up was a challenge.

NATIONAL TREASURE

After a quick polish the Spitfire was ready for John to take on a test flight. John was on the lookout for the slightest discrepancies in the way the aircraft handled. "When aeroplanes come apart to this degree, little things change," he explained. "The aileron trims might change, some of the engine parameters might have changed, you might pick up vibrations if the propeller is not right. So we're going to be flying it, settling it down, trimming it, basically getting it ready so when we leave it with Tom, we're happy and he's happy. I don't want to sign it off till it's perfect!"

Below Tom's Spitfire prepares for take off.

THE HEAVY-LIFT HELICOPTER

In 2009, when the European sailing team Alinghi had finished putting their new racing yacht, *Alinghi 5*, through initial sailing tests on Lake Geneva, Switzerland, they needed to find a safe way to transport the massive 90-foot (27 m) vessel to Genoa, Italy, for sea trials. They called in an MI-26 helicopter – the largest and most powerful helicopter currently in production – to airlift the 11-ton carbon-fiber catamaran to the open water. Conceived by Soviet engineers in the late 1970s as a military and civilian helicopter, the MI-26 can transport payloads weighing around 20 tons.

The team dismantled the catamaran's 17-story-tall mast for the journey and the boat took flight on the morning of 7 August 2009. The 170-mile (270 km) route passed down the Rhone Valley and across the Great St. Bernard Pass into Italy. The helicopter made a pit stop in Biella to top up its fuel tank, before continuing its journey to Genoa. The catamaran made a smooth touchdown later that afternoon in front of the Yacht Club Italiano. Her mast was then reattached and the craft was prepared for trials.

Over the years, the MI-26 has transported a wide range of cargo. In 2002 it rescued two Chinook helicopters stranded on mountains in Afghanistan. It was also used to move the carcass of a 20,000-year-old woolly mammoth found preserved in a block of ice in Siberia.

Above Tom's Spitfire in flight

As John went through the preflight checks, Mo watched anxiously from the hangar. "It's always a little bit nerve racking when you see the aircraft take off for the first time," said Mo. The Spitfire soared up into the air successfully and John spent the next few hours banking, looping and cruising above the clouds. As he returned to the airport, another aircraft touched down alongside him; it was Tom arriving back from business meetings in Washington, DC. After the engines cooled down, Tom walked over to greet John and shake his hand on a job well done. "I'm perfect now, I couldn't be happier! It looks terrific!" said Tom.

With a crowd of spectators at the hangar and the Spitfire ready to go, Tom prepped for his first flight, nervous with excitement. "One of the things that does concern me is, I would never want to be the American responsible for denting a Spitfire," he said. He climbed into the cockpit and with assistance from his ground crew, cleared the tarmac and fired her up. Spectators watched with anticipation. "To think back about the guys who were protecting Britain in this aeroplane and what they did, it's just unbelievable. The bravery and the skill they had, it's unparalleled." The Spitfire gained speed on the runway and soared up into the sky. With its D-Day livery visible on its belly, the glorious aircraft banked toward the horizon. It was a proud day for Mo and the team, and a childhood dream come true for Tom. "I don't think anyone really owns a Spitfire," said Tom. "I think you just take care of the Spitfire for the next person. Because it is truly a national treasure of Great Britain."

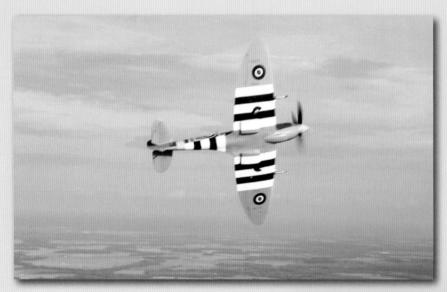

The Aircraft Restoration Company dismantled and flat packed Tom Blair's Spitfire to ship it across the Atlantic.

Clockwise from top left
The Spitfire in flight; Tom's Spitfire is fueled and primed for flight; the Spitfire Mk IX flies over the Maryland countryside; the Spitfire's Merlin engine.

Huge Halls

VARSITY HALL
THE HALL WITHIN A HALL

**Battling brutal blizzards and brittle brickwork, relocating the
University of Virginia's historic Varsity Hall was an immense engineering challenge
and trial of endurance for its move team.**

MOVE STATISTICS
Varsity Hall

Built	1857–58
Material	Masonry
Weight	500 tons
Length	36 ft. (11 m)
Width	52 ft. (16 m)
Distance moved	200 ft. (61 m)
Number of wheels	152

Today, Varsity Hall in Charlottesville, Virginia, may not look like an architectural wonder. But at the time of its construction back in 1858, it was one of the most technologically advanced buildings in America. Built on the University of Virginia campus, Varsity Hall was originally conceived as a student hospital. Set away from the main academic village, the hall was used to accommodate students quarantined with TB and other respiratory ailments.

The hall's architect, William A. Pratt, designed the building with a series of unique innovations intended to make recuperation as comfortable as possible. The hall had a unique heating and ventilation system, powered by a furnace in the basement, that pushed air through cavities in the walls to keep the atmosphere inside warm and constantly refreshed. The building was set at an angle on the site to catch the breeze, and its windows slid open into pockets within the walls to allow maximum ventilation. It even had indoor toilets, fed by cisterns in the loft. These features were novel back in the 1850s, making Varsity Hall one of America's earliest and most advanced student infirmaries.

RENOVATE AND RELOCATE

By the 1930s, Varsity Hall had been superseded and was converted into residential halls. But its unique features remained intact. So in 2005, when campus architects wanted to build a new teaching center on the land where it stood, rather than demolish the historic building, they elected to relocate it to a new site 200 feet (61 m) away. Moving Varsity Hall would cost over $1.5 million and was not a task for the fainthearted. The three-story building was built from more than 200,000 bricks and weighed over 500 tons. Its footprint was vast, concealing over

Above Varsity Hall prior to its relocation in 2005. The hall was built in 1857–58 and is one of the oldest student infirmaries in America.

6,000 square feet (500 m²) of floor space, and the path to the new site was a steep downhill run.

"A couple of movers backed out because they said the grades were too steep, they didn't have ways to hold the building back properly," said Maryland-based mansion mover Jerry Matyiko, striding on top of Varsity Hall's roof surveying the slope down to the new site. Jerry had moved many buildings in his time, but he had little idea of the hurdles ahead on this job. The features that made this building technologically unique would turn this move into a significant engineering challenge.

ONE LUMP OR TWO?

What sets moving a brick hall apart from moving a brick house is the difference in scale. The rooms in an average house are much smaller than those in a hall. Houses also generally have more partition walls inside. These act like the cardboard divisions in a supermarket wine-bottle holder to give the structure rigidity and help keep the exterior walls square. The rooms inside Varsity Hall were much larger. With fewer walls, the hall would be more vulnerable to twisting or collapse when moved.

Movers often get around this problem by slicing large buildings into smaller chunks to relocate them. Around the time of Varsity Hall's move, a team based in Georgia relocated the 150-year-old T.R.R. Cobb House 62 miles (100 km) across the state to Athens for preservation. The wooden building was too large to motor such a long distance intact. So they cut it into cubes for the journey. At the new site, cranes carefully lifted each section into place.

PUB CRAWL

Regulars at the Rob Roy in Victoria Park, Auckland, could have been forgiven for thinking that they'd had one drink too many when on 31 August 2010 the building began to slide off the berth it had occupied on the corner of Franklin Road for over 124 years. Built in 1886 along the city's original foreshore, the ornate hotel and bar was standing in the way of a major road-building project. The New Zealand Transport Agency wanted to build a $406 million tunnel directly underneath the building to remove a traffic bottleneck on Auckland's expressway network. Engineers needed to excavate the site directly beneath the Rob Roy to build the tunnel using "cut and cover" techniques.

To dig out the tunnel, but save the Rob Roy, engineers devised an elaborate plan to temporarily slide the building out of the construction zone, hauling it 144 feet (44 m) to a "holding site" where it would stay until the tunnel had been built and it was safe to move it back. The building's redbrick facade and decorative plasterwork were extremely fragile. To reinforce them during the move, the team drove steel rods through the facade. They sprayed a special form of concrete – called "shotcrete" – over its dilapidated rear walls to give them extra strength.

Two hydraulic rams positioned at the front of the Rob Roy gradually pushed the fragile structure along four concrete "runways" built along the length of the move route. Low-friction Teflon "pucks" positioned underneath the building enabled it to slide effortlessly along strips of stainless steel sealed into the tops of the concrete runways. It took two days to slowly push the pub to its temporary holding site. It made its return journey in April 2011 once work on the tunnel was completed, to the cheers of its locals.

Above D.T. Edwards' home relocated on rails by John Eichleay Jr.

Another option was to slide Varsity Hall on top of steel rails for optimal control. In 1914, John Eichleay Jr. used this method to relocate the magnificent home of publisher D.T. Edwards in Kinston, North Carolina. Hydraulic rams slowly pushed the building along steel tracks onto its new foundations, keeping the grand, three-story, neo-Georgian structure steady on its journey.

Jerry's team rejected both of these options. Cutting up Varsity Hall could cause irreparable damage to its unique features, and the slope to the new site was too steep to push the building on rails with any degree of control. So they opted to move it intact on a huge steel platform piggybacking on top of 152 wheels. This method would preserve the historic structure, but securing the fragile hall for its downhill run would pose a set of unique problems.

WALLS LIKE CHEESE

Work preparing Varsity Hall for its move began in the fall of 2004. "We're gonna separate this building right about the base of this window," said Jerry, pacing around the exterior of the hall dotting a thick black cut mark around its foundation. It was like watching a cosmetic surgeon plotting out incision marks on a patient's skin, but on a much grander scale. "From here up is gonna be moved. From here down is gonna be totally demolished after we pick the building up," he said, pointing above and below the line.

MOVING STACKS OF DOMINOES

By far the greatest challenge that Joe and Jerry had to overcome in moving Varsity Hall was its weak walls. The brickwork was brittle, the interior walls were only loosely attached to the exterior walls, and the areas around the windows were particularly vulnerable.

"We knew almost from the start that the windows slid up into the wall, which creates additional pockets," explained project manager Joe Jakubik. "Instead of having a box that we're moving, there's like breaks. It's kind of like moving stacks of dominoes."

The windows in Varsity Hall were unique. Traditional sash windows open and close within their framework. A concrete lintel usually supports the weight of the brickwork above. But Varsity Hall's windows were different. The two halves opened into the wall cavity to allow more ventilation inside the infirmary when the windows were wide open. There was no room for a brace over the window to support the wall above. So the walls and windows were like freestanding slabs that could easily fall down. To shore up these weak spots, Joe's team carefully removed all 40 windows, braced the openings with timber, and trussed the brickwork with wooden struts. "We've essentially taken that opening out of the picture," explained Joe.

To prevent the interior and exterior walls from falling away during the move, Joe's team ran steel cables through and around the building to pull the walls together. Over 2,500 feet (760 m) of cabling, tensioned tightly with 700 clamps, formed a giant steel spiderweb through the building. To avoid the tight cables slicing through the crumbling brickwork, they protected the corners of the building with metal guards. "As this is going down the hill, it is going to tip a little bit," explained Joe. "The bracing will prevent the walls from busting loose on us or pushing outwards."

Above left and right The windows of Varsity Hall slide up into a wall cavity, making the walls vulnerable during relocation.

Below left and right The exterior and interior of the hall were braced with steel cabling.

JEREMY PATTERSON: HEAVY HAULER

When an insurance company announced plans to build a new headquarters on the site of the historic Murillo Hall in Des Moines, Iowa, Rob McCammon from the local preservation society joined forces with a property developer to save the building. Their solution?

Call in larger-than-life mid-West heavy hauler Jeremy Patterson to move the 700-ton hall to a new site across town where it could be turned into luxury apartments.

Jeremy is a powerful force to be reckoned with. A former employee in a fast-food restaurant, he met his wife Tonya-Faye while flipping chicken. They left the heat of the kitchen behind to forge a new venture. In the space of just a few years, the Patterson family built up one of the largest heavy-haulage empires in the mid-West, shifting anything from massive mansions to vintage steamships.

Moving the Murillo would test even Jeremy's mettle. He had just four weeks to shift the 100-year-old hall, or it would be demolished. "The magnitude of the job is unbelievable," he said checking out the building. "It's one of the largest structures that we'll ever move, and one of the largest structures that will be moved. It's a feat that's almost impossible. But we're going to make it happen."

Jeremy's crew worked through the bitter winter of 2008 lugging over 100 tons of steel through the snow to build a frame to support the hall, connecting up 80 jacks to lift it and positioning 192 wheels to make it roadworthy. They waded through mud to wrestle the Murillo down the pavement onto the road aided by two 60-ton winch trucks lashed to its porch. "One of those cables breaks? It's going 500 feet [150 m]; it'll decapitate you," Jeremy growled, clearing spectators off the streets.

Yanking street signs out of the way, reinforcing the soft tarmac road with steel plates, and diverting traffic, Jeremy waged a heroic battle over the course of a day hauling the Murillo to its new site. "The whole time Mother Nature's been fighting us, kicking us in the ass. If it rains … we're set for doom," he said, racing to pull the building onto its new foundations before rain turned the site into a mud pit. He made it, just in time. "To the Murillo! Another 100 years, baby!" said Jeremy, taking a swig of celebratory champagne. "It's been one of them days!"

Jerry worked with engineers from International Chimney in Buffalo, New York, to carry out the move. Close inspection revealed that the hall was riddled with weak spots. "You have a building where just about everything that could go wrong is going wrong," said International Chimney's project manager, Joe Jakubik. Joe and Jerry had worked together many times before. It's awe inspiring watching this dynamic duo sizing up a building to move. Joe has an astute, almost X-ray-vision ability to scan the walls of any structure and immediately sense its weak spots, while Jerry can instantly tell you how many wheels he'll need and where they must go to mobilize any megastructure. When Jerry drives down the street in his truck, he imagines what the buildings he passes would look like rolling down the road. "That'd make a good-looking job," he'll say, as he drives past a redbrick, three-story, Victorian-style house with a proud turret. Neither Joe nor Jerry takes on moves that won't snap a good picture.

There was no doubt that Varsity Hall would be an impressive-looking move — that's if its brickwork survived the journey. "You could probably take this pocket knife here and take the whole brick wall down," said Jerry, scratching away at the soft red bricks in the basement walls. "These bricks are nothing. You could take this brick and put it in a bucket of water, and the water would bubble for about 15 minutes. It's like a sponge," he said as the bricks disintegrated in his hand.

The team's first task was to build a strong support frame made from long interlocking steel beams underneath the hall to keep it level as it moved. Jerry and Joe's men, donning waterproof gear and face guards, revved up their huge, diamond-tipped chainsaws and began cutting holes through the base of the walls for the beams. Water pumping over the saw blades mixed with the dry brick dust, splattering everywhere, quickly sullying the worksite.

Cutting through masonry walls 2 feet (60 cm) thick should have been hard work. But hacking holes through Varsity Hall's walls was like slicing through cheese. And the more holes they cut, the weaker the hall became. "We're cutting holes for these beams that run right through the entire building," said Joe as the saws hissed. "What that means is I'm cutting out about half the support that the building already has." To prevent the walls caving in above their heads, they braced each opening with wood and jacks. It took the team several weeks to thread the steel beams through over 100 holes to assemble the steel frame.

Below A huge steel frame constructed underneath Varsity Hall sits on wheels to support the building during the move.

157

BLIZZARD BATTLE

With such a sturdy frame in place, any regular house would now have been ready to lift off its footings. Not Varsity Hall. There were weaknesses hidden inside the structure that Joe and Jerry's team had to shore up before they could lift or move it. "You got this void here," complained Jerry as he uncovered another gaping hole in the wall. "To pick the building up with these voids, you have to fill them in," he sighed.

The voids were all part of Varsity Hall's ventilation system. The shafts were hard to trace and weakened the building. They weaved in and out of the two layers of brick that made up the exterior walls. Almost nothing held the two exterior walls together; a cavity ran all the way around the building. "The walls are hollow," explained Joe. "There's an air space in between. There's an inner width of brick, an air space, and then two outer widths of brick. It's kind of like a house within a house." The cavity meant that Joe and Jerry were in fact moving two separate structures at once; a hall inside a hall. If they were to lift both walls without tying them together, the steel frame could rip up through the weak bricks. To get around this the team had to fill the wall cavity with grout. This would glue the two walls together and provide a solid ring around the base of the hall for lifting.

Below With wooden struts bracing its window holes, and steel cables and beams bolstering its walls, Varsity Hall is secured for its downhill run.

HOTEL OPEN ... AND MOVING!

In 1869, movers in Boston, Massachusetts, launched a trailblazing attempt to move the hefty seven-story Hotel Pelham. City architects wanted to widen Boyleston Street in front of the hotel to cope with the increasing volume of traffic. They considered cutting a slice off the hotel to make way for a few extra lanes. But they eventually calculated that it would be cheaper to haul the hotel 13 feet 10 inches (4.2 m) back off the street to make way for traffic. With the hotel weighing over 5,000 tons, this would be one of the earliest-known attempts to move a large masonry structure.

Crews excavated underneath the building, sliding lengths of iron rails and more than 900 steel rollers under its walls and floors. They attached 72 screw jacks to the front of the building. By extending the jacks 1 inch (2.5 cm) at a time, they carefully pushed the hotel along the rails away from the road. The hotel's top speed was around 1 inch (2.5 cm) every five minutes. Incredibly, the hotel remained open throughout the three-month move. Plumbers fitted extendable water and gas pipes so that the guests could still check in and out while the hotel was in transit.

Upstairs, Joe's crew started pumping over 1,000 gallons (4,500 l) of grout into the cavity through holes drilled into the walls. The sloppy mix of cement and water dribbled down through the void and pooled around the feet of the men waiting patiently below to check that the grout reached the basement. "We got a bad leak. All stop!" cried one of the crew into his CB radio as the gray gunk streamed out of the wall. The bricks were riddled with tiny fissures, causing the grout to leak. The team had to plug each gap meticulously with rags to ensure that the cement ring set solid.

The work was grueling. The crew worked six-day weeks, bunking up in hotel rooms and apartments away from their families for most of the fall, preparing the hall for the move. Just as they plugged the last holes, a winter blizzard roared into town. Snow was the last thing Jerry and Joe needed. If the water in the grout froze before it set solid, the grout would lose its strength. They had to keep the hall warm. But it was full of holes. As the temperature dropped, they scrambled to wrap the bottom of the building in a tarpaulin to seal the openings and fired up six industrial heaters to raise the temperature indoors. "It's about 85 degrees [29°C]," said one of the crew, squinting to read the measure on a mangled thermometer. "You can see how hot it got by the way the thermometer melted!"

WHEN THE SHOW MUST MOVE ON

When American baseball team the Detroit Tigers needed a brand-new ballpark in downtown Detroit, Michigan, a huge obstacle stood in the path of their proposed field – the historic Gem Theatre. Built in 1927 as an addition to the Century Club auditorium next door, the two-tier brick show house in Spanish Revival style was a popular haunt for lovers of musical theater, films and cabaret and had been painstakingly restored in 1991.

The only way to save the protected and cherished local landmark was to relocate it to a new site. This would be no mean feat; the building weighed over 2,700 tons, had a sprawling footprint, and the proposed new plot was four city blocks away. The layout of the building also posed unique engineering problems. Not only were the Century Club and Gem Theatre inseparably attached, but the auditorium

and theater were both hollow structures, so had very little natural rigidity.

The stage was set for a truly showstopping move. Many movers joined forces under the auspices of engineering firm International Chimney, based in New York, and Expert House Movers, pooling equipment and manpower to shimmy the theater safely across Detroit. They braced the interior spaces with long steel beams and built a large metal platform underneath the structure to keep it level. It took more than 550 wheels to mobilize the hall for its 1,850-foot (564 m) journey. En route, the crew had to spin the theater around on its axis, so that its front entrance lined up with the street at the new site. The building's carefully choreographed dance across Detroit won the team the accolade of a Guinness World Record.

Left Varsity Hall descends the slope to its new footings balanced on 152 wheels.

BELTS AND SUSPENDERS

It took several weeks for the grout to set solid. But Jerry and Joe's battle with Varsity Hall was far from over. A tower on one side of the building was splitting away from the main structure. "There's nothing tying this tower in to this house," said Jerry on the roof, poking his fingers into the opening between the wing and the main wall. "The caulking gets wider and wider. It's like a big V." The tower was a later addition to the hall and poorly attached. If the hall tilted too far as it descended the slope, the tower could fall off. So they tied it to the main structure with tensioned steel cables. "You can see all our cables and our bracing to keep this from walking out," said Jerry inspecting the ties. "It's belts and suspenders."

After four months of arduous work, the day finally came to lift Varsity Hall off its old footings. "What you're going to have to watch when we start picking it up is how many of these bricks are going to fall down in between these beams because there's nothing bonding them together," said Jerry, signaling to his son Gabe Matyiko to keep an eye out for falling masonry. Gabe was busy hammering blocks of wood into the gaps between the beams and the bricks to prevent anything falling out and to ensure that all the beams lifted the structure simultaneously. "Say we just loosely wedge up and it's not really tight ... it'll cause one spot to pick before the other spot picks. When you do that, you start ... causing the walls to flex. With an old building like this that is brick or masonry, when they flex, they crack," explained Gabe, ramming more wood into a hole.

Jerry fired up the 12 hydraulic jacks underneath the rig. Spotters anxiously watched each jack to ensure that they lifted in unison. The tension was palpable. The months of cutting, shoring, grouting, strapping and bracing were about to be put to the test. Would Varsity Hall, with all its architectural anomalies, rise intact?

As the jacks slowly pushed up on the steel frame, a horizontal hairline crack gradually opened up around the base of the building along Jerry's thick black dotted line. Over the course of the afternoon, the entire building slowly rose 4 feet (1.2 m) into the air. "Outstanding," remarked Jerry with a cigar in his mouth, gleefully surveying the hovering structure. With adrenaline still pumping through his body, he picked up a huge sledgehammer and began to smash away the old footings, making space for the wheels. "Sledgehammer time!" he said with another crash. Jerry was in his element.

DOWNHILL RUN

"If something starts acting funny, starts going a little too fast, and somebody says 'chock 'em,' grab a block, set it in place," Jerry warned his crew on the morning of the big move a few weeks later. A streak of April sunlight lit up the 200-foot (61 m) pathway down the slope to the new plot. Crowds of students gathered to watch the move. "Watch the perimeter of the building, the brickwork is very loose. We don't want anybody struck by a brick," said Joe to his crew, unintentionally stoking the excitement of onlookers.

The team anchored bulldozers to the rear of the building with chains to stop the hall sliding down the hill. "Right ... here we go!" cheered Jerry as he fired up the

Below Varsity Hall hovers over its new foundations. Following the move, contractors built new walls up to the base of the building leaving space around the steel beams so that they could slide them out. Once the beams had been removed, the holes were patched up, creating a new basement underneath the main structure.

FAST FOOD

When the construction of a new shopping mall threatened to obliterate a pizza restaurant in Chattanooga, Tennessee, its owners went to extreme lengths to retain their customers and took away the takeaway to a new site. The owners called upon the mighty moving muscle of Jeff Kennedy, from Don Kennedy and Sons House Moving Co. in Huntsville, Alabama, to rescue the restaurant, which weighed 185 tons and soared 18 feet (5.5 m) high. It took the team two days to deliver the eatery 15 miles (24 km) through the center of Chattanooga to its new plot. En route, the firm delivered free pizzas for the move crew to keep their energy levels stoked up.

A larger restaurant relocation took place in Canada around the same time as the pizza restaurant's express delivery. Golfers at the popular Minnewasta Golf Course in Morden, Manitoba, were pinning their hopes on their local heavy hauler, Harold Minty, delivering a brand-new clubhouse for the beginning of their spring season. Their old building wasn't exactly up to par. "You have the roof collapsing ... a lot of the wood trim and stuff is falling off outside. We have walls shifting, the pipes are old and rusty," said the club's general manager, Chris Worley.

Rather than build a new clubhouse, they had decided to ship in a secondhand hall for a quarter of the price to replace their decrepit building. They had located a relatively new, large, wood-framed restaurant 23 miles (37 km) away that had recently closed down and was up for sale. Size rarely scares Harold Minty. But the vast timber-framed building, capped with a 50-ton vaulted roof, was too large to roll across rural roads intact. So the team sliced the hall into three sections, bracing its exposed ends with wooden props to prevent them caving in on what promised to be a bumpy ride.

Hauling a huge structure over a road glazed in thick ice might seem like a crazy idea. But there was method in Minty's apparent madness of waiting until the winter to drive the hall to the new site. The gravel roads along the route were riddled with natural springs, making them soft – not the best surface to sustain the weight of a heavy hall. But in the winter the springwater froze rock solid. The only danger was that in a Manitoba winter, temperatures plummet to minus 22 (-30°C) and the roads are covered in thick snow and slippery ice.

Harold knew this move would be a slippery schlep. And as they attempted to kick-start the trek and drive the three pieces of the hall onto the road, a winch line connecting one of the trucks to a section of the hall snapped as the wheels lost traction on the ice. The convoy ground to a halt. "It's not a convoy. The circus would be more like it!" joked Harold. With grit on the roads and determination in their hearts, Minty's team finally got the show on the road, and the convoy made headway across the frozen fields. "Hit the dusty trail, high gear wide open down the highway, get the heck out of Dodge," perked Minty's foreman Rick Mohoruk as the move got under way.

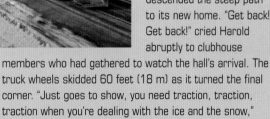

The journey took two days. Oncoming traffic faced a monster looming out of the mist as the crew battled through thick fog and bitterly cold winds. "It's slow," remarked Rick as tractors helped pull the loads up a particularly icy hill. "It's taking a little bit of time to get all three sections up." Things got dicey as the clubhouse descended the steep path to its new home. "Get back! Get back!" cried Harold abruptly to clubhouse members who had gathered to watch the hall's arrival. The truck wheels skidded 60 feet (18 m) as it turned the final corner. "Just goes to show, you need traction, traction, traction when you're dealing with the ice and the snow," remarked an onlooker.

"There's your Christmas present for the year!" said Harold as he successfully delivered Chris' new clubhouse. "Everything's down here in one piece, or in three pieces, as it were!" The reassembled building opened just in time for the golfing season.

engine on the side of the hall powering the wheels. While Jerry drove the wheels from outside, Gabe steered them underneath. "Any of them spinning?" shouted Jerry, checking that they had traction. The wheels turned slowly. Jerry dared not move faster than a few inches a minute. As the hall rolled off its concrete slab onto the gravel roadway, the lumbering load began to sink into the gravel. "All right, get all these flattened out," said Jerry, ordering the crew to lay down planks of wood in front of the wheels to spread out the load and provide extra grip. "It's rained the last couple of days," explained Gabe. "Usually this stuff, when it's really dry, packs down kind of hard as concrete. But it's a little wetter."

CHOCK IT! CHOCK IT!

They hauled the hall out of the hole, and leapfrogged the wooden boards in front of the wheels, creeping forward. It took three hours to move just 30 feet (9 m). The students, having called in take-out pizza, seemed transfixed by the slow-motion spectacle. Once the sun had dried out the gravel, Jerry cried, "We're gonna come off of this right now," gesturing toward the boards. "We'll be boarding all damn day!" He steered the hall off the roadway and

picked up the pace. But suddenly the building started to run too fast, dragging the heavy machinery behind with it. "Chock it, chock it, chock it!" called Jerry abruptly. The crew scrambled to throw blocks of wood in front of the 152 wheels to stop the hall running out of control. "The whole row went crazy!" said Jerry as the hall ground to a halt, referring to the wheels that had run too fast.

Soon, the hall began to creep forward again. This time, Jerry proceeded more cautiously, stopping every 2 feet (60 cm) for a pit stop to check the wheels. The farther down the slope they traveled, the steeper the incline became. An army of spotters surrounded the hall on the lookout for falling masonry. "It's a little disconcerting, you can see it leaning a bit," said Joe, eyeing up the building as it descended the hill. It took the entire afternoon to inch down the slope. They hung plumb bobs from the corners of the hall to ensure that the building lined up precisely with its new concrete slab. If they were a fraction of an inch out, the

Above Varsity Hall fully restored and repainted following its relocation

toilets wouldn't flush. "I'm on my mark!" shouted one of the spotters as they hit the drop zone. After a nerve-racking day, Varsity Hall finally reached the end of the road.

"I'm chocking it up now!" said Jerry as he switched off the engines, blocked up the wheels and slipped a celebratory cigar into his grinning mouth. "It's not going anywhere, anymore, nowhere, for 100 years," he said with relief. "Congratulations, that was slick, Jerry," said Joe as the pair shook hands. The move was a triumph and a testament to the weeks of arduous preparation and ingenious problem solving.

Over the coming months, the building was set down onto its new footings, the steel and bracing was removed, windows reinstated and the brickwork and interior extensively restored. Today, Varsity Hall sits proudly at the heart of the University of Virginia campus. Home to the Vice President for Research, this unique slice of 19th-century campus life has been saved for generations to come.

Varsity Hall's spongy bricks and cheese-like walls made sliding the building downhill onto new foundations an impressive feat of engineering.

Clockwise from top left
Varsity Hall hovers over its new footings; the team celebrate the move's successful completion; Varsity Hall restored and repainted in its new location; the crew guide the building into place over its new foundations.

Titanic Trains

THE TRAIN FROM BLOEMFONTEIN

Hauling a vintage steam locomotive from South Africa to Scotland turned into an epic 70-day adventure for a group of British steam enthusiasts sent on a unique mission to rescue the rusty relic.

MOVE STATISTICS

15F Class Locomotive

Built	1944–45
Weight	109 tons
Length	75 ft. (23 m)
Height	12 ft. (3.7 m)
Distance moved	6,000 miles (9,600 km)

The city of Bloemfontein in the heart of South Africa was once a major railroad hub. Steam engines plowed through the region every day transporting goods and people across the country. Today the city is where old steam engines go to die. The sheds at Bloemfontein train yard are stacked deep with engines destined for the scrap heap. Black smoke billows from the workshops as crews tear the rusty hulks apart with blowtorches and bulldozers, selling the metal for scrap.

The sight was a sad one for British steam enthusiasts Andrew Goodman and Jim Mitchell when they they arrived on a mission to rescue one of these relics in June 2007. Andrew had been seconded to Bloemfontein by the Riverside Museum in Glasgow, Scotland. The curators had bought one of the best-preserved engines in the yard, and wanted it transported to the museum, where it would be restored for exhibit.

Andrew has a huge passion for steam locomotives and had moved many before. He founded his haulage company, Moveright International, based in the West Midlands, England, following a personal quest to rescue and restore an old steam locomotive with the help of a group of friends. The project garnered much attention and galvanized Andrew's reputation as Britain's leading locomotive lugger, shuttling engines between yards, museums and tracks on a fleet of adapted trucks.

"I'm told as a babe in arms my parents took me to the Vale of Rheidol Railway in Wales, and I don't know whether it was the inhalation of the smells of the oil and the steam, but it got into one's body and it's been a thing ever since," said Andrew, inspecting the steam relics in Bloemfontein's yard. "The railway's not only an interest but a passion," he added.

Above The engine sheds at Bloemfontein train yard

LOST STEAM ICON

Protected from the heat and smoke under a canopy stood the engine that Andrew would spend the best part of eight weeks wrestling back to the United Kingdom: a North British Locomotive 15F Class engine and tender. Decommissioned in 1988, the majestic 60-year-old engine No. 3007 was one of the last surviving locomotives of its kind. It was built by the North British Locomotive Company in Glasgow between 1944 and 1945. The factory was the largest locomotive workshop in Europe in the early 1900s, shipping up to 400 locomotives a year all over the globe.

The 15F became one of the most common locomotives used in South Africa. Its powerful engine made it an ideal choice for hauling goods over the country's steep mountain ranges. The lucky 15F destined for a return trip to Glasgow had clocked up 40 years' service, hauling passengers and freight. It even pulled the prestigious Blue Train along the 1,000-mile (1,600 km) route from Pretoria to the Cape. "It became an icon for steam in South Africa," explained Jim, sizing up the hulking mass of steel. "It's really important to take one home, as much for the people of Glasgow themselves, because a good part of their heritage has been forgotten."

LUGGING THE LOCOMOTIVE

Moving the 50-foot (15 m) locomotive 6,000 miles (9,600 km) back to Glasgow would be a huge challenge since Andrew had never moved a locomotive this size such a long distance before. The team had reserved space in the hold of a cargo ship destined to dock in Durban, 400 miles (650 km) east of Bloemfontein, to take it back to the United Kingdom. Since the 15F's boilers were old and had not been fired up for over 20 years, they opted to load the locomotive onto the back of a flatbed truck and drive it to the port. They had seven days to reach Durban to catch the ship. This seemed like plenty of time, but Andrew suspected that the journey would not be easy. "There's always the unexpected, the unknown," he quipped. "We're not sure quite what the roads are like. It's not motorways, that's for sure!"

The team made haste on the first day, preparing to winch the locomotive up a ramp onto a truck fitted with special rails. Heavy chains would hold the wheels rock steady on the truck for the road journey. Jim surveyed the route out of the yard and realized that at 12 feet (nearly 4 m) tall the locomotive was riding too high to pass under the decorative gateway. "Too big!" he huffed as he wound up his tape measure. Workers armed with blowtorches raced to the rescue, slicing off the top of the gateway in seconds. With the locomotive and tender ready to load onto trucks for the journey, everything seemed to be going to plan. But when transport police arrived to inspect Andrew's superlong load, they immediately poured cold water on his plans. "We've issued the permit here, but there is a problem," said one of the officers, scrutinizing Andrew's loading plans. Although the main truck had six axles to distribute the locomotive's 109-ton weight across the road surface, the police were worried that the vehicle was not sturdy enough.

"It appears that the trailer that they've sent down, it wasn't really up to the job," said Andrew despairingly. He made frantic calls to try to locate a larger trailer. But such specialized equipment was a rarity in this part of the country; the nearest suitable vehicle was over 600 miles (1,000 km) away. The move was off and it looked as if the team would miss the boat.

Above The team remove the decorative entrance to the train yard so that the truck carrying the 15F steam locomotive can drive out.

CHANGE OF PLAN

Andrew spent most of the following day on the phone, speaking to the local rail operator Spoornet. The only way now that he could transport the locomotive to Durban in time to meet the ship was to shunt it along the rail network. But planning an unscheduled train journey at such short notice was a tall order. The fastest route to Durban was along the freight track. But the trains on this line traveled at 56 miles per hour (90 kph). If the 15F ran at this speed, its old bearings would probably seize up, leaving the leviathan blocking the line. So he was pinning his hopes on Spoornet granting him a track permit on a rarely used, rural route. The team could travel more slowly, but the track was very remote; if they broke down they could be stranded for days.

Above Winch cables connected to the rear of the 15F prepare to pull it up ramps onto a transport truck.

Andrew made his case to officials at the Spoornet offices. "As long as we can make Durban for Monday – that really is the critical day," pleaded Andrew to Spoornet's operations manager. It was Tuesday morning and the journey could take several days, so he needed a quick decision. "I don't want to make promises, but I'm really going to try my best," said the manager.

OIL HER UP!

Andrew spent the next few days running between his car camped in Spoornet's parking lot and their offices for meetings. When he wasn't discussing the minutiae of the logistics with Spoornet officials, he was sitting in his car with his phone pressed tight against his ear leveraging support from the project's backers in Glasgow.

Jim spent his time at Bloemfontein's train yard, galvanizing workers into action to prepare the 15F for one final train journey in the event that Spoornet gave the team the green light to proceed. There was lots of work to be done before the 15F, which had not run on rails for over 20 years, could safely ride on the network. "We

Below The yard team prepares to oil up the 15F locomotive for its rail journey.

can't rely on anything on the locomotive to play its part in a moving train," said Jim as they began the overhaul. "We're working frantically to make sure that absolutely everything is going to run smoothly." The yard crew scurried around the locomotive, meticulously greasing, polishing and brushing all of its connections and bearings. They sprayed all its brass components black to deter any thieves en route, as they had no idea where they would be stopping. It took more than 13 gallons (60 l) of oil to lubricate the moving parts.

THE RAILROAD THAT MOVED SHIPS

drew up designs for an interoceanic ship railroad: a 134-mile (216 km) railroad for oceangoing ships stretching from the Gulf of Mexico over the Isthmus of Tehuantepec in Mexico into the Pacific Ocean.

In Eads's scheme, ships would enter the mouth of the Coatzacoalcos River on the Atlantic side and be guided onto a huge sunken pontoon. The pontoon operated like a dry dock, built with ballast tanks that enabled it to sink or rise in the water when the tanks were filled or emptied. A large wheeled carriage sat on top of the pontoon with 50 rows of hydraulic rams built into its deck. Once the ship was lined up over the submerged carriage, the hydraulic rams extended upward to support its hull. With the ship balanced, the pontoon's ballast tanks emptied, raising the vessel high and dry. The carriage was coupled to three locomotives pulling in front of the ship and three engines pushing from behind. Running along three lines of track, the engines would pull ships weighing up to 12,000 tons across the land, delivering them from one ocean to the other.

In the second half of the 19th century, engineers across the world were drawing up schemes to enable ships to pass more easily between the Atlantic and Pacific Oceans to open up more direct trading routes. At this time, vessels had to navigate thousands of miles around the treacherous waters of Cape Horn to travel between the seas with their cargo. They urgently needed a maritime shortcut through Central America.

Alongside proposals to build interoceanic ship canals across Panama and Nicaragua, one novel scheme outlined plans for perhaps the greatest railroad system never built. In the 1880s, distinguished American engineer James Buchanan Eads

Eads had forged an esteemed reputation for himself over the years, having pioneered the construction of ironclad ships and built the first steel bridge across the Mississippi River. He maintained that the ship railroad would be far cheaper and less complex to build than the canals, and even proposed underwriting the $75 million cost of its construction himself to prove it was viable. Despite this, the scheme failed to win support from Congress and he died in 1887 without the idea ever being realized. Today the Panama Lock Canal provides the vital connection between the two oceans, while a far less impressive freight railroad runs across the Isthmus of Tehuantepec.

BIG BOY AND CENTENNIAL
GO OFF THE RAILS

When father-and-son team Dave and Bill Scribner were asked to haul two titans of the tracks off the rails and up to the top of a hill in downtown Omaha, Nebraska, they faced a formidable challenge. For these were no ordinary locomotives. "Big Boy," built in the 1940s and 1950s to haul freight over the Rocky Mountains, was the world's largest steam locomotive, while "Centennial," designed to mark the railroad's 100th anniversary, was the largest diesel locomotive ever built.

Omaha is the home of America's railroads. It was here in 1862 that Abraham Lincoln called for a railroad that would stretch from coast to coast. Today, Union Pacific Railroad controls more than 7,000 locomotives across a nationwide freight network from their Omaha base. The two restored engines, perched on a hilltop in Kenefick Park overlooking the city, would form a monument to these achievements.

It was impossible to build a temporary railroad through the busy downtown streets to transport these goliaths from the railroad yard to the park. So Dave and Bill had to lift the locomotives off the tracks onto special road-going rigs.

Worried that the roads might buckle under Big Boy's 420-ton weight, they spread its load across 144 tires. They attached two winch trucks to the front of the locomotive and a bulldozer pushing at its rear to wrestle the engine onto the road. The trucks flailed around in the air, losing traction as they pulled Big Boy over the pavement, reluctantly leaving the rail yard behind. With the roads temporarily closed, it was still a struggle steering the 98-foot (30 m) Centennial around street corners jammed with overhanging wires and traffic lights. Even on the straight and narrow, the loads were so wide that the crew had to clear some of the parked cars off the road to pass.

Crowds lined the streets to cheer the locomotives on their way. Everybody took a deep breath as bulldozers helped tug the engines up the steep slope to the park, fearful they might run away downhill. But eventually, the locomotives made it to the top. High on the hills over Omaha, today the giants sit as a permanent reminder of the glorious days of the transcontinental railroad.

They discovered that the locomotive's brake shoes were either worn out or not working. They had no spare parts to repair them, but the engine needed braking power or it might career out of control while traveling down one of the many steep mountain runs. To stop the 15F from running amok, they coupled 10 empty wagons to the rear of the train. Each wagon was fitted with a set of brakes capable of supplying 13 tons of braking force. Linked together, the wagons would provide enough might to bring the train to a halt.

Aided by local fixer Lizelle Wiid, who spoke Afrikaans, by Thursday afternoon Andrew had finally convinced Spoornet to organize a ride and route for the locomotive. "Just to be safe. What time are we leaving here?" quizzed Andrew, not quite believing his luck. "Six forty-five tomorrow morning," they confirmed. It had been an intense few days of negotiations. Lizelle was so overjoyed at the breakthrough that she leapt over and hugged the Spoornet manager. Spoornet were arranging a diesel-electric locomotive and driver to tow the 15F along the rural route to Durban. But they insisted that Andrew hire an additional driver to ride with them on board the old locomotive for safety. After a series of long calls, Andrew managed to locate Pieter Steenkamp, a seasoned 15F driver from Johannesburg, and booked him on an afternoon flight into Bloemfontein.

Later that day, the team coupled up the convoy for its eastbound trek. With the locomotive sandwiched between the wagons at the rear and the diesel-electric engine up front, the crew gave the 15F a final spit and polish. Their hands, arms and faces were slithery with oil. They had babysat the 15F well since its retirement. They would clearly be sad to see this unlikely survivor leaving its shed, and wanted it to return home in style. But just as Andrew was ready to retire for the evening, he received bad news: Pieter had missed his flight. "I mean, at the last minute! How on earth could anything further go wrong?!" he said as he got back on his phone to try to resolve the hitch. "All in the love of steam!" he sighed affectionately.

Above One of 10 empty wagons hooked up to the 15F convoy to provide additional braking power

Above With the diesel-electric train coupled to the front of the 15F steam locomotive, the convoy prepares to roll out of Bloemfontein toward Durban.

FINAL DEPARTURE

On Friday morning at dawn, the convoy was finally ready to depart. Warm sunlight streaked across the yard, lighting up the signature white stripe running down the length of the old engine. "I'm going to be honest with you, a few moments ago, it was quite emotional actually ... just watching everybody going around and checking," said Andrew. "When they loaded these locomotives onto the ship, who could ever have guessed that one would come back? It would have been beyond anybody's imagination that such a thing could have happened," added Jim.

Driver Pieter arrived in the nick of time to board the 15F. Just after 7 a.m., the locomotive and tender finally departed on their epic 6,000-mile (9,600 km) journey back home to Scotland. The yard crew waved a fond farewell as the engine creaked and groaned, rolling into the sunlight. "Nice steady run, now," said Andrew, cooped up in the 15F's driver cabin with Jim, Pieter and bags of provisions.

MORE GREASE!

They crept along the tracks with trepidation. The first few miles would be the most critical. Pieter feared that if they went too fast, the wheels could overheat and become damaged. Passing out of the city into the open countryside, the first 16-mile (25 km) leg took them over an hour. They pulled up at a secluded station in Sannaspos to check the locomotive's condition. Andrew knelt down and put his hands over a wheel bearing to take its temperature. "Getting a little warm," he said.

MIND THE GAP
MOVING A LONDON UNDERGROUND TRAIN

Following a major refurbishment of London Underground's Waterloo and City line in 2006, transport engineers faced a unique challenge getting the line's immaculate rolling stock onto its tracks. The Waterloo and City "Tube" line runs under the Thames and has only two stops, shuttling commuters between Bank Station in the heart of the City and intercity rail hub Waterloo Station. Getting trains onto the Tube lines is normally quite straightforward, via several access points above ground that feed into the underground network. With six lines converging at Bank Station, it might seem like an obvious place to transfer trains onto the tracks. But all the lines are at different depths with no linking tunnels between them. The Waterloo and City line is completely isolated from the rest of the Tube network – it is a tunnel sealed at both ends. So transport engineers had to adopt a different strategy and lower the driver cabs and carriages down a shaft 50 feet (15 m) deep that sits behind Waterloo Station.

Threading the 20 pristine carriages, each worth around $1.5 million, down a hole that was barely larger than they were, was a precarious operation. The team secured each carriage in a special lifting frame. A huge crane, extending 180 feet (55 m) above London, latched onto the frames with steel cables, and launched the carriages into the sky. The crane driver had to swing the carriages through a carefully choreographed arc to steer them past lamp posts, line them up over the hole and lower them onto the tracks. The crew had to down tools each time a strong gust of wind rushed through the street to avoid the carriages colliding with nearby buildings.

An unexpected challenge for the team was interpreting the unfathomably complex coding system assigned to the carriages. A five-digit number etched on either end of each carriage indicated the order in which they should be coupled up to form the train. When the team interpreted the codes incorrectly and mistakenly inserted a driver's unit instead of an ordinary carriage, they had to crane it out and start again because there was no space below ground to shuttle the carriages around. Eventually, all the carriages made it onto the tracks successfully and the Tube line was reopened.

THE TRAIN THAT PULLED A COURTHOUSE

In 1899, the residents of Box Butte County, Nebraska, witnessed a once-in-a-lifetime event: a railroad train pulling a huge building across the landscape. The reason behind this rail relocation was as strange as the sight itself. A decade earlier, when the railroad was built, its course had detoured around the town of Nonpariel, which at the time held the county seat. The tracks passed through the neighboring towns of Alliance and Hermingford, so the county held an election to determine which of these two communities should become the new county seat. Hermingford eventually won the honor and erected a new courthouse in the town. But the outcome of the vote was contested for many years, and in 1899 a rerun of the vote selected Alliance as the new county seat. Instead of building

a brand-new courthouse, Alliance's residents elected to make use of the existing building by buying it and moving it to their town.

In July 1899, engineers balanced the 54-foot (16 m) long and 40-foot (12 m) tall building onto the back of nine flat railroad wagons. They coupled two sets of wagons filled with coal to the front and rear of the building to help anchor the 100-ton structure to the tracks.

Ironically, the railroad that had originally instigated the relocation of the county seat then moved the county courthouse 9 miles (14 km) to Alliance. As the courthouse steamed along at 10 miles per hour (16 kph), a team of 75 people rallied ahead of it, widening the route and clearing obstacles out of the way to deliver the building intact.

Even at this slow pace, the bearings were heating up rapidly. On closer inspection, the team thought that old grease could be clogging up some of the bearings. Over the years that the locomotive had been in storage, dust and dirt had fused with the grease, forming a brittle mass. With no lubrication, the friction could generate enough heat to melt the insides of the bearings, causing the wheels to seize. The team forced sticks of solidified grease, resembling oversize strings of licorice, into every bearing on the locomotive. As the grease melted, it would flush out any solidified lubricant. They returned to the cab and signaled to the diesel driver to forge ahead, having agreed to stop every 16 miles (25 km) to re-grease the bearings. "You never know what is going to happen next, do you?!" quipped Andrew as they got back under way. "Just got to hope ... that we can maintain this speed, that nothing gets too hot, nothing causes us a problem."

The convoy made steady progress, passing through the towns of Ladybrand and Ficksburg en route to Bethlehem, arousing interest at every stop. Onlookers waved at the convoy as it passed through townships. At pit stops, inquisitive kids ran over to watch the crew, mesmerized by the sight of the lumbering behemoth.

It was rare for any train to travel on this line, let alone a vintage steam locomotive. The 15F gleamed majestically in the intense sunlight as it blazed a trail across the savanna. The journey took in stunning panoramas, passing through the golden-colored fields of the Sandstone Estate with the dramatic Maluti Mountains rising in the distance. They made good headway until Jim heard a worrying sound as the locomotive clattered along the tracks: "I am hearing a click. A rhythmic click on the left-hand side I haven't heard before." They signaled to the lead driver to pull in at the next stop.

Above The convoy en route to Durban

BROKEN BRAKES

"I'm concerned there's an old fracture on this brake assembly," said Jim, leaping out and examining the brake shoes on one of the wheels. "What worries me is if it catches and tears it." Part of the brake shoe was broken, and rubbing against the coupling rod. If they left it in this state, it could seize. But they had no tools to hand and the nearest town was miles away. "Well, there's not a lot we can do about it here is there!" said Andrew, as they congregated around the engine scratching their heads.

Jim devised a makeshift repair, jamming the brake shoe away from the wheel with a wooden wedge. "Chances are that will fall out!" he said, boarding the train. They set off again into the sunset. The train slowed down as it passed through a mainline station to allow platform crew to throw food and provisions into the 15F's cab. The night drew in and the convoy chugged into the mountains. "It's cold. It's freezing outside!" said Pieter as the temperature dropped abruptly. Having sweated all day, they now faced the prospect of shivering all night.

Above, left to right
Jim, Andrew and Pieter make a pit stop to inspect the steam locomotive's brake assembly.

179

THIS STATION IS NOW READY TO DEPART!

Sometimes it's not just a rail yard's rolling stock that leaves the tracks behind to embark on a road trip. Occasionally, the station buildings themselves need rolling across town. When the historically significant Hornsby Signal Box in Sydney, Australia, was blocking the path of a proposed new train line in 2007, heavy hauler Matthew Manifold of Mammoth Movers was called in to move the brick structure 400 feet (120 m) away from the tracks. Packed with old signaling equipment, the structure weighed over 300 tons.

In 2004, regular commuters in the desert railroad town of Deming, New Mexico, were left bemused as to where they should catch their train when their railroad station was moved to the other side of town. Deming's old station had arrived with the tracks in 1881. Everyone from cowboys to congressmen had passed through its doors. While the railroad company wanted to demolish the old building, locals were keen to preserve it, so called in father-and-son move crew Rick and Ricky Little to relocate the building to a park for preservation. The station was too long for Rick and Ricky to move intact, so they cut it into two smaller pieces for its trek across the desert.

Above The tender rides behind the 15F steam locomotive on its eastbound trek to Durban.

They wrapped themselves in extra layers of clothing and blankets and stoked newspaper, wood and coal into an old metal barrel, lighting a fire to keep them warm. "Make ourselves warm tonight, otherwise we will die in this place!" said Pieter as he rubbed his hands together. The bitter wind roared through the open cab as the convoy embarked on a roller-coaster midnight ride through the mountains. The team took turns to clamber into the empty tender car, cloaking themselves in blankets to try to catch some sleep as the train from Bloemfontein thundered on through the night. "I never thought I was going camping in South Africa when I came on this job!" joked Andrew. The conditions were hard going, but for this band of locomotive lovers, the trip was turning into the adventure of a lifetime.

BLOCKAGE ON THE LINE

It took two days and nights for the train to reach Durban docks. It emerged from a series of very long, pitch-black tunnels, screeching toward the end of the line well after nightfall. The team were shattered, but arrived back at the docks at first light to a new set of problems. They had arranged for a dock shunter to push the locomotive the last 2,600 feet (800 m) to the quayside so cranes could load it into the waiting ship. "There's a small technical problem. The diesel locomotive's broken," sighed Andrew, learning of the shunter's problem. It was two hours before it was fixed and the shunter began inching the engine ever closer to the water's edge. The old dock rails were rarely used and buried in grime and concrete. "They could derail, because the rails are so clogged with dirt at the front," said Jim, watching anxiously as the engine's wheels crunched over the grit. Then with just a few hundred yards to go, they hit another glitch. An abandoned truck sat blocking the line.

Above The 15F steam locomotive and tender loaded onto the deck of the floating crane that transferred them from the dockside into the hold of the *Diamond Land* cargo ship

With the cranes booked to load the locomotive into the ship that afternoon, and their shunter time running out, Andrew got on his phone and scrambled a team of dockworkers to clear the obstruction. The truck's battery was dead. They tried to start it with jump leads. But it was no good. "They've got 16 people here. Another four or five and we could just lift the truck out of the road!" said Jim wryly. Eventually, the operations manager signaled for a forklift and they shoved the broken-down vehicle out of the locomotive's path. "What can go wrong, will go wrong!" quipped Jim as they got back on the engine and rumbled toward the quayside.

Later that afternoon, with the locomotive gleaming in the sunlight, a huge floating crane sailed to the dockside and lifted it onto its deck. "It's not very often we get to lift locos like this!" said the crane's operator with a glint in his eye. As the sun set, the 15F left South African soil behind and made it safely into the hold of the *Diamond Land* cargo ship. The crew lashed its wheels to the ship floor with chains. "The Captain was saying there's 5-metre [16 ft.] swells forecast for going round the Cape. So it'll need to be well tied down," said Jim, crossing his fingers as the ship's deck folded down and the locomotive disappeared from view. At 8 a.m. the following morning, the *Diamond Land* set sail.

JOURNEY OF A LIFETIME

Four weeks after leaving Durban, the locomotive finally arrived at Immingham Docks on the River Humber, England, where it was unloaded onto a trailer for the last 300-mile (500 km) leg of its journey to Glasgow. Jim looked on proudly as the locomotive emerged out of the hold unscathed. "It's quite a special day really, because it's the first time the locomotive has been on British soil for 60 years. It's a nice moment," he said. Truck driver Ben Swift measured the lofty load before setting off. "The

THE TRAIN FROM BLOEMFONTEIN

Saved from the torches of the Breaker's Yard,
Thirty years of service under African guard.
Restored and oiled for one last ride,
Six thousand miles to take in its stride

CHORUS
*This is the Train from Bloemfontein,
Mighty workhorse of the African Plain.
This is the Train from Bloemfontein,
Off to Scotland, home again.*

Into the valley of a thousand peaks,
Twisting and turning, the Mountain Class creaks.
Icy wind bites, but no blazing coal,
This arduous journey is taking its toll.

Forty days and nights in the open sea,
Out on a passage spanning oceans, three.
Stormy waters brew around the treacherous cape,
Will the precious cargo safely escape?

For old times' sake a final parade,
Under shadows of spires and colonnades.
Conquered the ocean and the open road,
It's the end of the line for this majestic load.

CHORUS
*This is the Train from Bloemfontein,
Mighty workhorse of the African Plain.
This is the Train from Bloemfontein,
Safe in Scotland, home again.*

bridge heights on the motorway are 16 foot 6 [503 cm]," he explained. Riding 16 feet 3 inches (496 cm) high, the 15F would just squeeze through. "Good enough!" said Ben, jumping into the cab.

The locomotive blazed a trail across the sunny British countryside. Steam enthusiasts who had been tracking the 15F's journey on websites flocked to bridges, highway banks and gas stations en route to snap its picture. It rolled into George Square in Glasgow under the veil of night for display over the weekend. The following morning, crowds amassed around the vintage locomotive, surprised at its sudden appearance and blissfully unaware of its epic adventure. Jim and Andrew shook hands, exhausted, but proud of their monumental achievement. "Well, it was a bit of a journey of a lifetime wasn't it? I mean, to do what we did was incredible," said Andrew with a smile.

Over the next three years, the locomotive was restored for permanent display at Glasgow's Riverside Museum. A commemorative song called "The Train from Bloemfontein," written by composer Daniel Pemberton to celebrate the locomotive's journey, received a premiere performance at Glasgow's Clyde Auditorium sung by the local ACE Chorus. The locomotive's heroic rescue has made it one of the most famous icons in the city, a fitting monument to Glasgow's lost locomotive-building heritage.

Above The 15F steam locomotive and tender on display in George Square, Glasgow

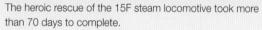

The heroic rescue of the 15F steam locomotive took more than 70 days to complete.

Clockwise from top left The 15F locomotive in Bloemfontein's engine yard; engineers lift the locomotive out of the cargo ship; the engine is transferred onto a truck at Immingham Docks for its journey to Glasgow; the crew inspect the 15F convoy at a pit stop; the Bloemfontein train-yard crew.

Colossal Collectors

THE MAN WHO COLLECTED A TOWN

As a young boy, Adam Woolley had a dream – to build his own historical Texan town. It would be 40 years in the making, as he herded up more than 20 historic buildings from across the state to create a slice of rural life from a bygone age. The journey of one of the buildings would be particularly poignant: a period stone mansion that was a gift for his mom.

MOVE STATISTICS	
The Spillman House	
Built	1866
Material	Limestone
Weight	230 tons
Length	40 ft. (12 m)
Width	20 ft. (6 m)
Height	35 ft. (11 m)
Distance moved	3,300 ft. (1,000 m)
Number of wheels	58

Adam Woolley's heritage is firmly rooted in Texas. His great-grandfather was born in 1884 in Industry, which was the first town established in the state by immigrants from Germany. Five generations of Adam's family have grown up in the area. Although little remains of the original town today, its history holds a special place in Adam's heart. "My family has been in Texas for about 150 years," explains Adam, who was brought up in Austin, Texas. "Texans think of themselves as their own republic, and tend to hold on tightly to their heritage."

As a child, Adam spent a lot of his time in Industry with his great-grandfather. Over the years, he cultivated a fascination for the town's architecture, falling in love with its wooden-clad picture-postcard buildings. His great-grandfather passed away when Adam was 9 years old. As he left no will, most of his assets reverted to the State of Texas and the family lost the historic building that had been their ancestral home. Just one strip of land remained in their possession. It was a moment that would change Adam's life for ever. He vowed to re-create the look and feel of Industry one day when he was older. He even wrote an essay in the fourth grade about his grand vision. Building his town became his obsession.

As he grew up, Adam spent much of his pocket money collecting "props" that he would one day put in his town. He did odd jobs in exchange for old car license plates, gas pumps and signs that he amassed in the family garage. Adam knew that he needed a career that would allow him to build his town both physically and financially. So when he was 18 years old, he enrolled to do a Bachelor of Science degree in Architectural Engineering at the University of Texas. As he began to earn a regular wage designing office blocks and corporate buildings, the size of the props he could buy grew in scale. He soon found himself in possession of bar counters

and old store fixtures. His collection, now stored in a much larger warehouse, grew so vast that he was able to turn it into a business. He began making sets for feature films, leasing out his props to furnish the scenery. Over the subsequent 19 years, he worked on over 500 productions.

FIELD OF DREAMS

In 1998 Adam bought a 31-acre (13 ha) plot of land in Beecave, just outside Austin. It was here that he would finally assemble an array of antique buildings and his collectables to create his dream town. Adam spent the next three years clearing the brush off the site to establish a blank canvas on which he could create his masterpiece. He drew up the plans for Main Street in close collaboration with two production designers he had befriended while working in the film industry – Emmy nominee Cary White and John Frick. Together they plotted out the orientation of the street to maximize the natural light inside the buildings and obscure any views of modern developments nearby.

Adam began herding buildings onto Main Street in 2001. The first structure to ride into town was a Rosenwold Schoolhouse, threatened with demolition in Seagoville, Texas. Julius Rosenwold was a philanthropist who donated millions of dollars in the early 1900s to build education centers for African Americans excluded from state schools by segregation. His schools became heritage landmarks and were placed on the list of the nation's most endangered buildings by the National Trust for Historic Preservation in 2002. Adam used a local mover to help him jack up and truck the schoolhouse to Main Street. With the first building in place, he christened the site Star Hill Ranch.

It wasn't long before Adam's neighbors were tipping him off about other buildings in the area that were available for transportation to his town for

Above and below The buildings Adam has collected and moved to Star Hill Ranch

preservation. Most were donated to him on the condition that he covered the costs of moving the buildings. He quickly acquired a post office, a butcher's shop and then a chapel. As the buildings pulled up in Main Street, he restored them and furnished them with props from his collection. Adam's town was slowly coming together and attracting visitors from across the state. It had not always been easy. But with the support of his loving and dedicated family, who had always encouraged and believed in him, Adam's childhood dream was slowly turning into a reality. "It's sort of like the 'Field of Dreams.' You build it, people will come," he recounts.

By 2008, Adam had relocated more than 22 buildings to the site. Star Hill Ranch was used regularly to host weddings and events and also served as a shooting location for several movies. Adam converted the cottages into guesthouses where visitors to the town could stay overnight. His church was leased out every weekend and had a congregation of 60 worshippers. Adam's town was no dusty museum piece. It was a living, breathing space for the community to cherish. Star Hill Ranch was where condemned buildings were given a new lease of life: a slice of lost American history preserved for the community.

LAST PIECE OF THE JIGSAW

In 2008, Adam decided to add one final, very special building to Main Street to complete his town: a new home for his beloved mom, Judy. She had lived in a house in the countryside for over 43 years. But the city had steadily been encroaching. "Now it's all houses, lots of traffic and people driving too fast down the road and I want to just have a less complicated life," she explained. Judy's dream was to live out her retirement at Star Hill Ranch. "This town is so beautiful. The historic buildings, the flowers, the smells ... I love it all and I can't wait to move out here!" she said excitedly.

There were several smaller buildings on Main Street that Judy could move into, but none was quite right. So Adam went searching for the perfect mansion for his mom. Eventually, he found the Spillman House: a limestone mansion built just after the Civil War in 1866. Shrouded by century-old oak and elm trees, the ornate two-story building was once at the heart of a cotton plantation. When Adam found it, the old house was abandoned and run down. But he immediately fell in love with its character. "The first time I saw this place, I felt this immediate connection to it. You know when you see it that it is something special. It's a real jewel. This is right up my mom's alley!" he said, exploring its spacious rooms.

Adam purchased the property and set about working out how to move it to Star Hill Ranch. The

Below The Spillman House prior to relocation, shrouded by old oak and elm trees

HOW TO MOVE A STONE MANSION

Moving the Spillman House was a precarious operation. The limestone walls of the mansion were extremely soft and the lime mortar in between the stones was crumbling. There was next to nothing holding the structure together. To support the mansion for its journey, Harold's team built a sturdy platform underneath the house made from long interlocking steel beams.

To prevent the stonework at the base of the house from disintegrating as they lifted the structure off its foundations, the team had to build a safety net around the bottom of the walls. The net was knitted from steel bands. To secure it in place, first they drilled 200 small holes at regular intervals around the perimeter of the mansion through the bottom layer of stone. Then they threaded lengths of steel banding through the holes, wrapping it tight around steel beams braced to the stonework farther up the walls.

"Banding is a bit like stitching a big quilt with a big needle," said Bill, one of Harold's workers, as he guided the long steel threads through the holes.

It took the team two weeks of patient needlework to thread over 3,300 feet (1,000 m) of steel through all the holes. In some areas, where the limestone was particularly weak, they doubled up the banding to provide extra support. Once in place, the bands were tightened to pull the stonework together. They could tell whether the bands were tight enough by the "twang" the loops made when hit with a hand tool. "Basically that's how we judge how snug we are getting with the banding," said Bill plucking the bands with a wrench. The bands resonated with a high-pitched sound. "When they start getting to a certain point they all start carrying a tune. It's almost like picking a guitar. Hey, that sounds pretty good!" he said as he plucked the bands down the length of the wall.

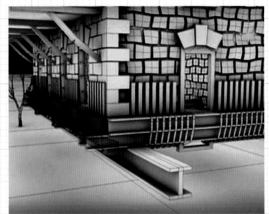

Spillman Estate was only about ½ mile (under 1 km) from Main Street, but to reach the hilltop town Adam would have to haul the fragile building over some wild Texan terrain. The mansion was colossal and its crumbling walls looked like they might give way at any moment. Weighing in at 230 tons, the house was far heavier than anything Adam had moved before. So he called in Texas' leading ranch wranglers, father-and-son team Harold and Gary Snow.

MOVING THE CROWN JEWELS

The Snow family has been moving buildings in Texas and around the country since 1942. Harold Snow's first moving job was at the age of 12. He trained as a foreman under his father and traveled the United States picking up moving techniques from other movers and building up the family business. His son, Gary Snow, joined the family firm in 1973. With the Snows having over 60 years of experience and a wealth of expert knowledge, Adam decided that there was no better team he could entrust with the crown jewel of Star Hill Ranch.

As soon as Harold set eyes on the building, he knew that it would be a challenging job. The walls of the Spillman mansion were built from limestone 16 inches (40 cm) thick and the lime mortar in between the stones was crumbling. "You could actually pinch a bit of lime mortar from between the stones and mash it to nothing really," he said. The building was sitting on top of solid rock, just a few inches from the surface of the ground. This would make it difficult for the team to dig out its foundations. The modern extensions on the side of the building were also loosely attached to the main structure and unlikely to survive the move intact.

Where to begin? Taking into consideration the integrity of the building and the state of the mortar, one option was to totally dismantle the structure and move it stone by stone. But it would be a painstakingly slow process and the deeds to the house stipulated that the building could not be taken apart. So Harold's team had to move the mansion intact. Buying and moving the building would cost Adam $400,000, making the project his most ambitious to date. He was not taking it lightly. "It's really intimidating. I wake up in the night wondering if I've lost my mind," he said, helping the crew clear up the debris inside the house, preparing it for the road.

Below Adam Woolley inspects the banding holding the brittle stonework in place around the base of the Spillman House.

READY FOR LAUNCH

The crew built a support frame made from bright-red steel beams underneath the house to keep the building level as they moved it, and braced its weak walls with steel bands. The Texan heat was intense and they worked with a constant threat of being bitten by rattlesnakes. The workers scoured the worksite regularly

Above A Diamondback rattlesnake on the loose in the grounds of the Spillman House

to remove the deadly serpents. "Oh! We got one. Nice Diamondback!" shrieked crew member Matt, as he used a stick to retrieve one of the slithery intruders from the estate's well. The Western Diamondback rattlesnake is America's deadliest snake and one bite could kill a man in five hours. "The house is coming right through this area. Come here, big guy!" Matt said as he enticed the snake into a container to move it off site. "Put him in the bucket, and get him to a new home!"

With the mansion braced and bolstered, Adam and Judy came to watch Harold's crew lift it off its old footings and slide it onto wheels. "It's very exciting and very nerve racking. I feel like we are getting ready to launch the space shuttle here!" beamed Adam excitedly. The crowd waited with bated breath. After what felt like an eternity the building started to lift up. "We got a crack, she's breaking loose," called one of the workers as the wall separated from its footings. The team had achieved liftoff. Over the course of the day, 19 powerful lifting jacks raised the Spillman House 3 feet (1 m) into the air, giving them just enough space to position 58 wheels underneath. "I am so excited I can hardly stand it. Where's our champagne?" said an overjoyed Judy, inspecting the mansion floating in the air in front of her.

MOVING HOME

Move day dawned overcast. Adam rushed to greet the movers with some bad news: storms were forecast later in the day. "I was just listening to the weather report and it looks like we have a pretty strong Canadian front on the way down that is going to kick off some pretty severe weather – some rain and hail," he reported. This could have caused problems for Harold's team, who had to haul the mansion across a field to reach the road to Star Hill. After weeks of men working around the house, the ground had been churned up and it would rapidly turn into a quagmire if it rained.

Below Hoses link the hydraulic jacks positioned underneath the Spillman House to a jacking machine that raises the building up off its old footings.

They needed to move the mansion off the soil and onto the tarmac road before the heavens opened, or the heavy building would sink into the mud.

Despite the prospect of rain, Harold felt confident that the house would survive the journey intact. Before they left the site, he placed a bottle of water on the top of the 35-foot (11 m) chimney. It was a 40-year-old Snow tradition. If the bottle remained upright for the duration of the move, then they knew that the job had gone perfectly.

THE HOUSE WITH THREE HOMES

Those who caught sight of a small stone cottage trundling happily through the Missouri countryside on the back of a bright-red truck back in September 2005 could be forgiven for asking why the building was moving or where it was moving to. But the more perceptive onlooker might well have also queried why the stones on its facade were inscribed with numbers.

For the cottage blazing a trail through the woods that day, it was not the first time that it had been uprooted and hauled to a new home. This was the second big outing for the charming historic building, once home to American pioneer Squire Boone. Squire built the cottage from rock rubble, and used it as a base camp while laying traps and hunting along the Missouri River. The cottage was originally totally dismantled and relocated stone by stone in 1987. The movers marked a distinct number on the front of each

stone to ensure the stones could be reinstated in their original position at the new site. It took the team over a year to dismantle, move and rebuild the cottage. But they made an indelible error; when they numbered the bricks they used permanent ink!

When Missouri movers John and Joe Matyiko of Expert House Movers were called in to relocate the building a second time in 2005, there was no messing around with jigsaw plans. The Matyikos, who pride themselves on shifting buildings wholesale, jacked it up and trucked the cottage to its new home intact. It took just a day to drive the numbered cottage to the picturesque Boonesfield Village in Defiance, Missouri, where it was positioned alongside a collection of other period buildings. A quick sandblast removed all trace of the numbers on its stonework, as the cottage settled into its third home.

TRAPPED IN THE WOODS

The first obstacles that they had to contend with were the oak and elm trees that blocked the truck's exit. "We got trees on the front, on the back, and both sides and they are all historic trees," explained Josh, one of Harold's workers. "We are trying to preserve all of them and essentially, the house is trapped." Prohibited from chopping down the trees to clear a passage to drive the house forward, the team attached winch trucks to its rear to pivot it around so they could reverse the building through a small clearing in the bushes. The winch cables screeched as they tightened, swinging the house around at a snail's pace. Harold's crew watched intently to ensure that none of the branches collided with the mansion's weak walls. "It seems tense," said Adam watching the house creep under the trees. "It seems like the labor-and-delivery room right before childbirth!"

It took four hours to turn the house around ready to cross the field. By this time, the light was fading. But the team had to plow on. "We're doing this in the dark, but we have to do it!" explained Gary as they set off across the field. Despite their determination, they were too late, and the heavens opened. "The rain is here but it ain't bad yet," said Edgar, one of the workers. Adam quickly hooked work lights to the back of his 4x4 to light a path for the team across the field. Battling the sodden ground, the building inched across the field. But the drizzle soon turned into a heavy downpour.

"Hold it! Hold it! Hold it," cried Harold suddenly. One side of the mansion was sinking into the ground. If the house leaned any farther, the walls could crack or give way. The team made a quick assessment. "As long as we can get over the hump and it don't get too slick then we will be just fine," said Bill, one of the crew. They cabled an extra winch truck to the front of the convoy and slowly heaved the mansion out of the hole. They battled through the mud and finally reached the road, exhausted, 17 hours after the move began. "All right men, god darn thank the lord it's up here," sighed Harold with relief as they secured the truck and downed tools for the evening.

Right The team use winch trucks to pull the rear of the house around into a clearing in the trees before backing it out of its old setting.

THE SHIP THEY SAILED ON ICE

It's as if someone pulled the plug on the ocean that once lapped at the keel of the mighty SS *Ticonderoga* steamship. Perched in the center of an immaculately manicured lawn, the vessel preserved high and dry at the Shelburne Museum in Vermont is an awesome spectacle.

Built in 1906, the SS *Ticonderoga* once cruised Lake Champlain, shuttling passengers between Vermont and New York. The iconic steamer was retired in the 1950s. Rather than break up the old ship for scrap, antique-building collector Electra Webb proposed moving her inland to become the centerpiece exhibit at Shelburne, sitting alongside a lighthouse, schoolhouse, bridge and several homes she had already moved to the site.

The task of sailing the 892-ton hulk over 2 miles (3 km) of boggy farmland to reach the museum was "truly gigantic in every respect," according to Webb. To create a firm path for the ship, the movers waited until the swampy ground froze solid and built a temporary 300-foot (92 m) rail track stretching inland from the water's edge. Special cradles, shaped to hug the ship's lower hull, would hold the

vessel steady as she rolled along the rails balanced on top of 16 railroad wagons.

In November 1954, tugboats steered the ship into a huge basin built in Shelburne harbor. With the SS *Ticonderoga* floating over the submerged cradles, the team drained the water to leave the ship resting upright on her supports.

On 31 January 1955, the ship embarked on an epic 65-day voyage across the frozen bog. Winch trucks pulled the ship 250 feet (76 m) per day along the tracks. As she moved forward, workers raced to uproot the track at the rear of the ship and leapfrog it ahead of the vessel. The weather was so cold that the crew had to use burning torches to stop the lubricating grease on the tracks from freezing.

Bulldozers were called in to rescue the ship when an unusually early thaw left her marooned in sludge in the middle of a dairy farm. The expedition soon got back on course, and the SS *Ticonderoga* sailed past cows and even traversed a working railroad line to reach her majestic meadow mooring, where she was restored to her former splendor.

ROLLIN' HOME

After a night's well-earned rest, Harold's team woke up to the welcome sight of blue skies. The storm had passed. Now the drive up the hill to Star Hill Ranch could begin. "Lord willing ... she's just about ready to get up and get herself a new address!" said Harold, pleased to get the house onto the tarmac. "Let's rock and roll!" quipped Bill, as the wheels hit the road.

Neighbors lined the streets to witness the spectacle. "You sure don't see something like this around here very often," said a policeman who had been drafted in to close off the highway to allow the house to pass. To calm the fears of any startled observers who weren't expecting to see a house rolling down the road, Harold's crew were wearing overalls emblazoned with the phrase, "Don't worry, I do this all the time!" on their backs. Adam watched, praying that the gift for his mom would arrive in one piece. The house turned into the ranch and crept slowly onto its new foundations. Despite the wind, rain and rough terrain, 3,300 feet (1,000 m) on from the building's original home, the Snows' trademark water bottle was still standing upright on top of the chimney. "We're home ... home as it gets," said Bill, relieved that the move had been a success.

Below A bluegrass band gather under a tree en route to serenade the Spillman House on its way to Star Hill Ranch with a specially composed song, "Rollin' Home."

ROLLIN' HOME

Down in a valley where live oak trees grow,
An old house lies sheltered in a white cotton grove.
For five generations it stood mighty tall,
Now its future's threatened by a 10-ton wrecking ball.

CHORUS
Rollin', rollin', this house is rollin' home,
Across the hills of Texas rolls 200 tons of stone.
Rollin', rollin', rollin' down the road,
One hundred years of history on an extra wide load.

One brave man believes that this house can be saved,
His mother needs a mansion to live out her last days.
So upwards it's lifted on great beams of steel,
And now it's heading homewards on 30 sets of wheels.

Slowly, so slowly, this huge house moves along.
Anxious, nervous workers checking bad things don't go wrong.
The rock is soft and aged, and the thick walls are unsound,
And any slight mistake could bring it crashing to the ground.

ALL WRAPPED UP

No present would be complete without its wrapping, and under the cover of darkness Adam arranged one final tribute. The old Spillman House was gift wrapped with over 200 feet (60 m) of bright-red fabric; a beautiful bow, 6 feet (2 m) wide, adorned the mansion's facade. It isn't every day that you get to give your mom a new home, but when you do, it should be gift wrapped. Adam asked Judy to close her eyes and walked her over to unveil his surprise. "Have a look!" he said to his mom, standing proudly in front of the house holding her hand. "Oh, that is wonderful!" gasped Judy as she opened her eyes. "Oh my gosh. It's just unbelievable. What a wonderful surprise!"

Before Judy could move in, Adam had three years of painstaking restoration work to carry out on the house to return it to its former splendor. Judy was finally able to call the Spillman House her home in the summer of 2011. Adam's childhood obsession is now his legacy, and while he has no immediate plans to move any more buildings, the Spillman House may not be the last building to pull into the gates of Star Hill Ranch ...

Left Adam unveils his present to his mother, Judy.

199

Harold and Gary's team battled rain and mud to transport Adam's mother's dream house to the historic Texan town he had created.

Clockwise from top left
The move team; the Spillman House gift wrapped for Adam's mom, Judy; the Spillman House secured to its new foundations, repainted and restored; aerial photograph showing the Spillman House at the new site; the team on the move.

Tall Towers

TOWER POWER
MOVING HEATHROW AIRPORT'S
CONTROL TOWER

On a cold fall night in 2004, a mysterious vehicle taxied along the runway at London's Heathrow Airport. This was no plane. The soaring beacon of light was a brand-new air traffic control tower embarking on a remarkable journey.

MOVE STATISTICS	
Heathrow Airport Control Tower	
Built	2004–2007
Material	Glass and steel
Weight	1,000 tons
Height	105 ft. (32 m)
Distance moved	1¼ miles (2 km)
Number of wheels	144

London's Heathrow Airport is the world's busiest international airport. Over 67 million passengers and 1.3 million tons of cargo pass through its gates every year. With air traffic forecast to double over the next two decades, in 2002 construction work began on a brand-new terminal building. With an enormous free-span roof soaring over 130 feet (40 m) high and a fully automated baggage system, Terminal Five was designed to feed an extra 30 million passengers through Heathrow every year. Its 60 additional gates would dramatically increase the number of planes taxiing on the far side of the airport. The only problem was that it would be impossible for air traffic controllers to see this zone clearly from Heathrow's existing air traffic control tower; Terminal Five's high roof would obscure their view.

The airport urgently needed a much taller tower, to enable its air traffic controllers to safely monitor and guide all the extra aircraft. For the best view, the new tower needed to be positioned in the center of the airport, next to Terminal Three. But this posed a conundrum for the tower's construction team. The sweet spot for the tower sat at one of the busiest crossroads in the airport, surrounded by operational gates and taxiways for baggage trucks and shuttle buses. Turning the bustling junction into a construction site would cause chaos. So the construction team came up with an ingenious solution. They opted to build the main hub of the new control tower outside the airfield perimeter, then move it into place.

Moving the cone-shaped structure, built from steel and glass, just over a mile (about 2 km) from its construction site onto its new footings would be a unique engineering challenge. "We're preassembling the tower ... which is 32 metres [105 ft.] tall, weighs approximately 1,000 tons, and we are going to transport it

vertically across the runway to its final destination at Terminal Three, which may seem like a really mad idea!" explained project manager Nick Featherstone, inspecting the gleaming structure as workers installed the final panes of glass. "The structure itself is inherently top heavy. It looks rather like a wine glass; a lot of its weight is at the top half," he explained. To stabilize the structure, the team were erecting a huge triangular support frame around the base of the tower. Made from heavy steel girders, the temporary frame broadened the base of the tower, lowering its center of gravity to prevent it from falling over during the transportation.

Once the frame was assembled, the team planned to lift the 1,000-ton monster onto multi-wheeled transporters to drive it across the airport. To reach the footings, they would have to guide the tower across one of Heathrow's runways. With over 100 aircraft an hour streaming in and out of the airport, it was impossible to shut down the runway to allow them to complete the move during the day. Instead, the airport authorities had given Nick's team a razor-thin window of just five hours in the dead of night to guide the tower into position.

Until the day of the move, a row of bolts encircling the base of the tower – imaginatively called "holding-down bolts" – anchored it to a temporary foundation in the ground. "The whole of the cab is supported on holding-down bolts," explained Mike Wade, the man in charge of the move. "Once we've got the trailers under each corner and supporting it we can actually remove these nuts completely and then we'll lift the complete unit." Once the tower reached its final destination Mike's crew could simply lock it into position with an identical set of bolts.

RIGHT ON THE NIGHT

It took months of planning for the team to assess every possible risk. More than 5,000 square feet (450 m²) of glass encircled the tower's observation area, making it terrific for visibility, but extremely fragile. The team met regularly to run through each disaster scenario and prepare a response should their worst fears be realized. "We're on the move, we're going down the ramp and there's a whole load of creaking and groaning out of the structure," said construction manager Peter Czwartos, posing one such scenario to the gathered team.

Below Heathrow Airport's new control tower, designed to sit at the end of a pier at Terminal Three

Left Cranes lift the roof of the control tower into position at the construction site, which was situated outside the airfield perimeter.

"Stop. All stop," responded Mike thoughtfully. "We try and determine where the noise is coming from."

"I mean, presumably this thing's likely to creak and groan anyway but it's trying to ascertain what's normal and what's beyond the norm, I suppose, isn't it?" said Nick as they set about classifying what distinguished a bad noise from a very bad noise.

The ramifications of any holdups or calamities on move night were far reaching. If the runway was blocked or damaged, it could wreak havoc on world flight patterns and jeopardize Heathrow's business. "If we were to close part of the runway just for a day you're probably talking about some very big figures, certainly £30 million [$48 million] plus," explained Nick. "That, as a business risk, would be unacceptable because it's not just the £30 million loss of revenue. It's also the share price and the bad PR that we would get from the airlines and everything else. We've got to make sure it goes right on the night, simple as that."

MADE IN ITALY

Several weeks before the move took place, a unique shipment arrived at Heathrow. Gleaming red like a fleet of pristine Ferraris, the three transporters that would propel the tower across the airport were delivered for testing and installation underneath the tower. "All singing, all dancing, very nice!" quipped one of the workers as the transporters' Italian operators put them through their paces.

The machines – known as self-propelled modular transporters (SPMTs) – were developed in Italy by transport specialists Fagioli and are used to transport bridges, oil tanks and other loads too wide or too long for traditional trucks to handle. The solid flat base of each trailer sits on top of 48 rubber-tired wheels. Each wheel can rotate independently in a full circle, making them ideal for precision parking a building. The wheels are attached to the trailer by huge hydraulic arms which move up and down automatically as the wheels pass over bumps and dips in the road to keep the base of the trailer level. With an average operating speed of around 3 miles per hour (5 kph), Fagioli's SPMTs are renowned for providing one of the slowest and smoothest rides in the world.

For particularly long loads, the trailers can be attached together to create a centipede-like train of wheels. For Heathrow's control tower, one trailer would sit beneath each corner of the triangular support frame built around the building's base. Huge metal pins would lock the base of the tower to the transporters. The transporters' 144 wheels would also spread the load of the control tower – weighing twice as much as a jumbo jet – over a wide area to protect the concrete runway from damage.

The trailers are powered by a diesel engine and operated by a remote-control console attached to the main unit by a long umbilical cord. The 1,000-ton load would be driven by just a single operator. Fagioli engineer Roberto Radicella was the man chosen to pilot this unique craft. "It is like a PlayStation!" he said as he pushed and pulled the console's joystick, testing the wheels. "It's very, very simple!" he added, as the tower's engineering team looked on in envy. "Fantastic. I just won't ever drive it!" said one of the British workers, disappointed he wouldn't get the chance to take the wheels for a spin.

OMINOUS FORECAST

The week of the move brought a worrying forecast. "This is the bit that I don't like: moderate to high lightning risk," said Mike as he analyzed the weather data for the week. There was a high risk of thunderstorms rolling in on move night, bringing with them the possibility of lightning strikes. On the

Below A steel support frame assembled around the base of the tower spreads its weight over a larger area and hooks directly into the SPMTs positioned underneath.

THE LOST WINDMILL OF DENMARK

marking each miniature piece with the number of its supersize counterpart to help the carpenters in Iowa rebuild it. The "flat-pack" windmill and the scale model were loaded onto a truck, driven to the Danish coast and shipped to New York. They arrived in December 1975, encrusted white with ocean salt. The 66-foot (20 m) sails were too long to travel intact on the U.S. highway. So the team had to slice them in half to drive them to Elk Horn.

Over 300 volunteers helped the local carpenters reconstruct the windmill at Elk Horn. The metric measurements inscribed on the plans caused some confusion. So the team had to make multiple runs to check the model windmill on display in Main Street. Cranes eventually hoisted the windmill's massive sails back into place, and on 15 March 1977, Elk Horn Windmill was complete, ready to grind its first batch of corn. To the horror of the carpenters, there were some pieces left over! Luckily, these turned out to be spare parts that the Danish carpenters had included in the shipment in case anything got damaged in transit.

When American Dane Harvey Sornson went on a trip to the land of his ancestors, he wanted to return home to Elk Horn, Iowa, with a keepsake. The memento Harvey had in mind was no natty knickknack; it was a supersize souvenir ... a full-size, working windmill!

Mesmerized by wooden windmills since his childhood, in spring 1975 Harvey proposed that Elk Horn pool their resources to transplant a Danish windmill to their village as a monument to their heritage. Settled in the mid-1800s by Danish immigrants, the village of Elk Horn in rural Iowa has one of the largest concentrations of Danes in the United States. Many responded, "You're crazy!" But within a short time of outlining his bold plan to move a disused mill from Norre Sneue in Jutland to the village, Harvey had raised around $30,000 in donations to help fund the cost of dismantling, shipping and rebuilding the mammoth mill.

In October 1975 Danish carpenters began dismantling the 127-year-old windmill. To ensure that each piece of the jigsaw – from its long sails to its heavy stone mills and gears – fitted back together in the right place, they meticulously numbered each component. They even built a 6-foot (2 m) replica of the 60-foot (18 m) building,

THE TOWER KEEPERS

In the early part of the 20th century, thousands of wooden grain elevators dotted Canada's rural landscape as the country's grain industry boomed. Built alongside railroad lines, these "cathedrals of the prairies" were used to weigh, sort, grade and store grain before it was loaded into railroad cars and transported to ports and cities. Today very few grain elevators survive. Many were bulldozed or burned to the ground as farming technology evolved. But Saskatchewan farmers John and Joy Weir have attempted to save several of these antique towers by moving them to their farmstead near the town of Perdue.

Moving the towers, which can soar up to 200 feet (60 m) high, is a tall order. The Weirs' first attempt to move an elevator to their farm ended in catastrophe; as they drove the tower off the road onto their land, the wheels of the transport truck skidded on ice, causing the tower to topple (see below). The following year, they tried again, working with local structural mover Jake Weib to transplant an elevator from the town of Kinley. This time the move was a success (see right). Although the red paintwork on its slatted wood paneling is faded, this lost piece of Canadian farm heritage still functions to this day.

morning of move day, the team would have to undo the holding-down bolts and lift the structure off the ground in preparation for the evening rollout. From the moment the team released the tower from the ground, it would not be earthed. With few tall buildings within the vicinity of the airport, the tall tower posed a risk for the move team; it could be like walking around the airport attached to a lightning rod. "Lightning's a worry because the structure, with this height and being steel, is prone to lightning strikes," explained Mike. "If it does strike while it's sitting on the trailers it could come through the driver units and damage them or worse, it could injure or kill one of the operators if they're anywhere near."

Above The plan was for the tower to glide across the airfield on three SPMT units positioned under each corner of the support frame.

PLOTTING A COURSE

The team monitored the weather closely in the days leading up to move night, scheduled for 28 October. Postponing or canceling the operation at short notice would be a difficult call for Nick to make. They had worked closely with many different airlines, cargo transport companies and subcontractors to ensure no planes, buses, trucks or other equipment would be left parked overnight in the path of the tower. It was a huge logistical operation.

It wasn't just large aircraft that could bring the move to a halt either. The smallest obstacles threatened to become major hazards. Although the route to the new site was largely flat, there were tiny bumps and gaps between the concrete slabs forming the airport taxiways. Even though each of these gaps was only a fraction of an inch in size, the gaps could derail the whole operation if they were not carefully negotiated.

To plot out the optimum route for the tower, the move team walked the course several times in the middle of the night when the runways were closed. Wielding torches and tape measures, they meticulously mapped out the position of every tiny obstacle. The team calculated that the tower needed four major turns to complete the journey so that on arrival the base of the tower would align very closely to its new foundations. The flat, wide-open sea of concrete was devoid of any distinct landmarks. So to ensure that the tower turned at the right junctions and avoided potential hazards, they plotted out its route using GPS. "By using the GPS we can actually find the points on the night to determine where we stop," explained Mike.

COLOSSAL CARGO

On the morning of the move, there was a scurry of activity in and around the control tower. Cleaners worked their way down the circular staircase in the tower's base sweeping up any leftover detritus. Workers used industrial tape to secure loose items such as fire extinguishers to the walls and floor. Anything falling off the tower could be a danger to the move crew. Since the tower would be passing into the airport's operational zone, it had to undergo the same security checks as any other cargo or passenger. Two sniffer dogs called Mini and Spud gave the tower a thorough inspection. "It's moving into a restricted area where everything has to be thoroughly checked," said their minder, eyeing up the colossal piece of cargo. "This is just another item, although a rather large item. It's got to be thoroughly checked before it goes through."

By 10 a.m. the tower had been cleared for takeoff. But inside the "war room," the team were unsure whether it was safe to do the move tonight. Nick had received weather reports forecasting that strong winds and storms could be passing through the area later in the evening, posing the risk of lightning. He had to decide whether to lift up the tower this afternoon. Once the holding-down bolts were released, there was no going back; they had to move the tower to its new footings and fasten it to the ground.

Nick kept the crews on standby for a decision until 5 p.m., waiting for updated forecasts. Eventually, he gathered the team into an emergency meeting. With no sign of an improving forecast, he decided to postpone the move. It wasn't a decision taken lightly and came as a blow to Heathrow's management. Rescheduling the move to another day would be a major logistics exercise involving contacting airlines and operators within Heathrow to get them to agree on a new time and date.

MOVE DAY TAKE TWO

The following day, despite the forecast of more unsettled weather ahead, the team gathered for a second attempt. "There's still a moderate chance of lightning today, that's the forecast. But we expect it to be over by about lunchtime," said Mike, ushering his team about as they frantically undid the holding-down bolts. To prevent another holdup, they installed a temporary earth line to ground the tower in the event of a lightning strike. "If we do experience a lightning storm when we've backed off and we're sitting waiting for the go tonight, it means if we do get struck by lightning it'll come through here rather than damaging the trailers," explained Mike as the team fastened the line to the tower. With all their permissions in place, the team slowly lifted up the tower, transferring its hefty weight onto the crawlers. A thousand tons of glass and steel were now in Roberto's hands.

The sun set, leaving the team with the clear skies they had hoped for. Concerned that the slightest interruption might wreck Roberto's concentration, no crew were allowed near the tower on its journey. So Mike, Nick, Peter and the rest of the team sat waiting patiently inside cars and trucks a safe distance away. With the transporters' engines whirring and Roberto's hands primed on the joystick, the team were given the green light to begin the move toward the runway at 10 p.m. "Prepare to invade!" radioed Peter, after receiving the all clear from air

Below Heathrow Airport's new air traffic control tower lit up for its journey across the airfield

THE TOWERS THAT MOVED OVER A MOUNTAIN

Around the time of Heathrow Airport's ambitious control-tower move, a team in Manitoba, Canada, were attempting an equally daring relocation. But their tools and technique were a stark contrast to those used in London. Eschewing reams of math, Harold Minty's crew relied on brute strength, grit and determination to haul five colossal grain towers – taller than London's Nelson's Column – across a mountain sheeted in ice. The towers, used to store and process grain, belonged to a grain plant in the sleepy valley of La Riviere that had gone out of business. A grain company in the town of Somerset, 20 miles (32 km) away, had bought the towers and were relying on Harold's crew to shift them to their plant. "It may be amazing to the normal public but for us, we do it on a regular basis every day," said Harold with aplomb as his crew rubbed lubricating soap onto steel beams under the towers to slide them onto 180 wheels for the journey.

Built from wood and steel, the towers had a combined weight of over 500 tons. Work began on the move in fall 2004. But it proved a formidable challenge to wrestle the towers onto the road, and it was winter before they were ready for their trek. With temperatures as low as minus 22 (-30°C), thick sheets of ice and around 4 feet (over 1 m) of snow coating the roads, Harold's team fearlessly dragged the lumbering giants out of the valley across a mountain. Crowds sipping hot coffee inside cars lined the route to watch the spectacle. "This is the best entertainment that money could buy!" said one excited onlooker.

Like the team at Heathrow, Harold was using wheels that could adjust up and down to compensate for bumps in the road. Unlike the team at Heathrow, his crew had to make every adjustment by hand. It was a laborious process and it took an entire day to travel the first ¼ mile (400 m). Halfway up the mountain, the wheels underneath the tallest tower lost their grip and began to skid. The convoy screeched to a halt. But it wasn't long before tractors and bulldozers thundered across the white hills to the rescue. Clouds of black smoke bellowed out of the trucks and the towers leaned precariously as the convoy groaned up the steep hill; from a distance, the metal monsters resembled oversize Daleks invading a frozen planet. After 20 hours of struggling through sleet and snow they finally reached the summit. "Not worried for my life any more!" quipped one of the workers.

They picked up the pace, slogging 16-hour days to guide the towers across the icy prairies. Bridges groaned and tires occasionally burst. Wildlife even took roost in one of their trucks. "We've been on this job so long and on the road for so long now we've got pigeons riding with us!" joked Harold, with a bird fluttering on his truck's dashboard. After eight days on the road, the towers received a warm welcome in Somerset as they emerged from the fog. The prospect of new jobs brought the locals out onto the streets cheering. "It's just like a big sleeping giant coming through a small little quiet town," said an onlooker. "I've got a delivery here for you!" said Harold with relief as he pulled the convoy into Dan Stevens' grain yard. The team connected the new towers and the equipment was up and running in time for the harvest.

THE LEANING TOWER OF KIMMERIDGE BAY

took close-up photographs of the tower's surface. Using a computer to piece the photographs together, they created a detailed two-dimensional image showing the arrangement of the stonework in close-up detail.

Work on the move began in June 2006. Engineers erected scaffolding around the tower to help them dismantle its 16,272 stones into crates. Conditions on the exposed cliff were brutal, with winds gusting up to 70 miles per hour (113 kph). Deconstructing the tower took nine months. The team then used the photogrammetric plan to reunite each stone with its neighbor at the new site, while the grid ensured that the spaces between the pieces of masonry were all faithfully reproduced. The last stone was laid into position on 25 February 2008, the ambitious project safeguarding Clavell Tower for at least another 150 years.

Perched high on the edge of a crumbling cliff in the middle of the stunning Dorset countryside in England is Clavell Tower. Built in 1830 by the Reverend John Richards, Clavell Tower served as a watchtower and folly. The three-story stone landmark was also used by Thomas Hardy as a frontispiece for his *Wessex Poems* and inspired P.D. James' novel *The Black Tower*. But over the years, the cliff – rising 330 feet (100 m) above sea level – eroded away and by 2000 the tower sat just 13 feet (4 m) away from collapsing into the sea. Forging ties with heritage group the Landmark Trust, locals launched an urgent campaign and raised over $1.2 million to save the iconic tower.

The building was in a dilapidated state, so the options for rescuing it were limited. There was no guarantee that shoring up the cliff to combat further erosion would work, while underpinning the tower with a reinforced foundation could leave it hanging in thin air as the cliff crumbled away beneath it. So engineers decided to move the tower to a more stable site 82 feet (25 m) inland.

As the tower was too brittle to move intact, engineers had to dismantle it, moving it one stone at a time. This was a formidable challenge given that the tower was a listed monument. English Heritage agreed to preserve the building's listed status on condition that the engineers re-erected the tower with every stone put back in its original position. To ensure that the tower didn't go all Humpty Dumpty, engineers used a technique called photogrammetry to map out the precise positions of the stones before they took the tower apart. They marked a grid across the tower's masonry, numbered each individual stone and then

traffic controllers. There was now no turning back. They had to reach the runway by 11 p.m. It was an emotional moment for the team, who had spent many months planning for this unique event. "I think I might cry!" said Peter as the tower slowly moved off into the night.

Above The control tower is maneuvered into place over its footings near Terminal Three.

With colorful lights illuminating the top of the glass tower, it blazed a stately trail past the stands and parked aircraft. The transporters' wheels slowly inched up and down over the joins in the concrete, keeping the tower perfectly level on its journey. From a distance, driver Roberto was barely perceptible against the towering beacon. The sight of the tower taxiing across the airport was a disconcerting spectacle for the tower's design team, who were watching from the sidelines with their eyes wide open with awe. "It's the first time I've seen one of my buildings disappearing off into the distance like that!" said one designer. "Just watching those transporters go off down that ramp there. The front ones pick up, the rear ones come down, it just keeps the entire mast vertical. It's a wonderful bit of technology, fabulous stuff."

CROSSING THE RUNWAY

The team made it to the runway on time. A Qantas aircraft was still taxiing down the runway. So they waited nervously on one side waiting to cross, knowing that in just over five hours' time, the airport would need to be back up to peak capacity with

Above The Heathrow tower move crew, including Nick Featherstone (second from right), celebrate their successful delivery.

one plane leaving every minute. "We certainly don't want to be leaving this thing anywhere near the runway tonight, not if I want a job tomorrow anyway!" quipped Nick, anxious to get a move on.

At 11.15 p.m. they finally received the all clear. Roberto inched the tower onto the runway as the rest of the crew took in a deep breath. An army of spotters marched behind the tower wielding brooms and wheeled garbage bins. Spotlights on the backs of trucks beamed light onto the surface of the runway to allow them to scrub every inch of pathway behind the tower clean. A stray screw could cause devastating damage to a plane on takeoff, so it was vital that they ensured that the tower left no debris in its wake. The tower completed its graceful glide across the runway by midnight and Roberto carefully rotated it through three precise turns to line it up with its new footings. "Must be about another 50 yards, 70 yards [46 m, 64 m] ...," calculated Nick, peering out of his car as the tower inched closer to its gate.

The crew drove around the airport to greet the tower as it docked into place near Terminal Three. Amazingly, after traveling 1¼ miles (over 2 km), its final resting position was accurate to within a fraction of an inch. It was a huge success and a tribute to the months of careful planning that the crew had put into the night's events. "We just moved a 1,000-ton object within 2 or 3 millimeters [3/32 in.]! Fantastic, what else can you ask for!" cheered one of the workers with a huge grin on his face. The crew shook hands and cheered, giving themselves a much deserved pat on the back. "Well, we might try and get one last cool lager just to send ourselves to sleep nicely tonight. Looking forward to a good night's sleep!" said Nick, relieved the move had been completed without a hitch.

Over the following weeks, the control cab unit was jacked up to its final height of 285 feet (87 m), giving air traffic controllers unparalleled 360-degree visibility around the entire airport. Full operations were passed from the old tower to the new tower in February 2007. With the opening of Terminal Five the following March, London Heathrow had a brand-new skyline, and a major upgrade to help it cope with 21st-century air travel.

Opposite Once the tower had been moved into place over its new foundations, jacks raised the structure to its final height. The "build, move and jack up" method of construction avoided the need to use cranes to build the structure from the ground up in situ at Terminal Three, which would have interfered with ground traffic.

Special transporters were imported from Italy to maneuver Heathrow Airport's new air traffic control tower into place with the minimum of disruption to take-offs and landings.

Clockwise from top left
The tower under construction outside the airfield perimeter; jacks raise the tower up to its final height; the tower gives air traffic controllers unparalleled 360-degree visibility around the entire airport.

Index

Acknowledgments

This book – and the television series it accompanies – would not have been possible without the help and creative talent of a huge team of people.

I am indebted to the Matyiko Family for introducing me to the wonderful world of structural moving, in particular Jerry, Joan, Gabe and Sarah Matyiko, who have been a constant source of encouragement. Thanks are also due to Rick Lohr and Joe Jakubik of International Chimney Corporation for helping me to get to grips with the engineering principles involved in structural moving; Eugene Brymer of the International Association of Structural Movers for introducing me to the community of structural movers; and all the members of the IASM and other transport specialists who have been so cooperative and generous with their time, allowing us to film their impressive projects. I am also grateful for the cooperation of all of the individuals, families, groups and communities who own the buildings, vehicles, and machines we have featured.

I'd like to thank the editorial and design teams at Quercus Books, in particular Richard Green and Emma Heyworth-Dunn for their enthusiasm and guidance with this project. I am also grateful to Robert Gwyn Palmer at DCD Media for helping us to get the book off the ground; Stuart Williams for making so many constructive suggestions for improving the text; and John Driftmier, Cherry Dorrett, Jamie Lochhead, William Lorimer, Lee Reading and Leesa Rumley for their written contributions to chapters in this book.

At Windfall Films special thanks are due to David Dugan, who, as the series' Executive Producer, has offered myself and the whole production team unflagging enthusiasm and support over the years; Head of Production Birte Pedersen, who has contributed so much to the book and television series; the team of dedicated production staff who have worked on the series, including Simon Barker, Duncan Bulling, Stephen Bonser, Nick Carew, Jody Collins, Mike Davidson, Steve Dorrett, Roeland Doust, Matthew Fletcher, Anne-France Bos, Susan Harvard, Jason Hendriksen, Tyson Hepburn, Ash Jenkins, Jessica Jones, Jayne Johnston, Kristina Obradovic, Alex Ranken, Chris Roots, Nick Scullard, Paul Shepard, Ian Strang, Bettina Talbot, Marissa Verazzo, Gwyn Williams and Keith Wilson; Musician Daniel Pemberton for his entertaining sounds and songs; and Dave Throssell's team of animators at Fluid Pictures who created the distinctive CGI for the series and illustrative 'blueprints' for this book. I owe additional thanks to Lee Reading, whose meticulous research and organization has helped pool together the stories and imagery featured in this book.

I am grateful to the following production executives and commissioning editors for supporting the series over the years: Dan Chambers, Justine Kershaw, Robi Dutta and Andrew O'Connell at Channel Five; Ann Harbron at Discovery Canada; and Bridget Whalen, Sydney Suissa and Janet Vissering at National Geographic Channel. I would also like to thank all of my friends for their support on this project, including Jeanette Goulbourn and James Taylor.

Finally, thank you to my loving family Angela, David and Francesca Massarella, for their inspiration, encouragement and unconditional support.

Picture Credits

A FIREFLY BOOK

Published by Firefly Books Ltd. 2011

First printing

Publisher Cataloging-in-Publication Data (U.S.)

Massarella, Carlo.
Monster moves : adventures moving the world's biggest structures /
Carlo Massarella.
[224] p. : col. photos. ;   cm.
Includes index.
Summary: The incredible feats of engineering and ingenuity to move the world's biggest structures.
ISBN-13: 978-1-55407-931-5
1. Moving of buildings, bridges, etc. II. Title.
690.837 dc22   TH153.M377   2011

Library and Archives Canada Cataloguing in Publication

Massarella, Carlo
    Monster moves : adventures moving the world's
biggest structures / Carlo Massarella.
Includes index.
ISBN 978-1-55407-931-5
    1. Moving of buildings, bridges, etc. I. Title.
II. Title: Adventures in moving the impossible.

Published in the United States by
Firefly Books (U.S.) Inc.
P.O. Box 1338, Ellicott Station
Buffalo, New York 14205

Published in Canada by
Firefly Books Ltd.
66 Leek Crescent
Richmond Hill, Ontario L4B 1H1

Printed in China

Developed by:
Quercus Publishing Plc
21 Bloomsbury Square
London
WC1A 2NS